IntechOpen

Supporting open minds since 2005

Noise and Environment
http://dx.doi.org/10.5772/intechopen.87872
Edited by Daniela Siano and Alice Elizabeth González

Contributors
Juan Rios, Celedonio Aguilar, Jānis Indulis Dundurs, Inka Janna Janssen, Mia Suhanek, Sanja Grubesa, Vijaya Baskar Veeriayan, Rajendran V, Jiho Lee, Himanshu Dehra, Mohamed Abdelsabour Abdelsabour Fahmy

First published in London, United Kingdom, 2021 by IntechOpen
IntechOpen is the global imprint of INTECHOPEN LIMITED, registered in England and Wales, registration number: 11086078, 5 Princes Gate Court, London, SW7 2QJ, United Kingdom
Printed in Croatia

British Library Cataloguing-in-Publication Data
A catalogue record for this book is available from the British Library

Additional hard and PDF copies can be obtained from orders@intechopen.com

Noise and Environment
Edited by Daniela Siano and Alice Elizabeth González
p. cm.
Print ISBN 978-1-83968-239-1
Online ISBN 978-1-83968-240-7
eBook (PDF) ISBN 978-1-83968-241-4

For EU product safety concerns:
IN TECH d.o.o., Prolaz Marije Krucifikse Kozulić 3, 51000 Rijeka, Croatia, info@intechopen.com or visit our website at intechopen.com.

Meet the editors

Daniela Siano was born in Naples - Italy, and graduated in Aeronautical Engineering from the University of Naples "Federico II", Italy in 1994. Until 2001, she was a researcher in the Acoustic and Vibration Department at C.I.R.A. (Italian Aerospace Research Center). From 2001 until now, she has been a researcher at the National Research Council of Italy (CNR) in the field of acoustic and vibration in the transport field. She is responsible for the Acoustic and Vibration Laboratory. She is qualified to University Associate Professor – Sector Machines and Systems for Energy and Environment, and is also qualified in the Engineer Profession. She had a scholarship granted by M.A.R.S.- Microgravity Advanced Research and Support - Dornier GmbH (Friedrichshafen - Germany), Euromaster in "Total Quality Management" at DIMP – Department of Mechanical and Production Engineering – University of Naples Federico II – Italy, as well as assistant in "Fluid Machines I", "Fluid Machines II", and "Applied Mechanics". She is the tutor of more than 30 Master's students and 7 Ph.D. students. She is an expert evaluator within the EU 6th and 7th Framework Research Programme, in Transport-Aeronautics in 2006 and 2007. She is a project expert evaluator in the Ministry of Economic Development, Italy. She is a referee for different international journals and session organizers collaborating with SAE conferences. She is the author of more than 100 scientific papers published in international journals and conferences proceedings and editor of two scientific books: 'Noise Control, Reduction and Cancellation Solutions in Engineering', ISBN 978-953-307-918-9, 308 pages, DOI: 10.5772/1375, and 'Fuel Injection', ISBN 978-953-307-116-9, 262 pages, DOI: 10.5772/55402.

Alice Elizabeth González is a full-time Professor at the Environmental Engineering Department, Faculty of Engineering, Universidad de la República (UdelaR, Uruguay). She currently teaches undergraduate and postgraduate programs and oversees graduate and postgraduate students. She also leads research, development, and applied research projects. She is a Researcher of the National Researchers System of the National Research and Innovation Agency. Her main research topic is noise pollution, including environmental and urban acoustics, noise and vibration control, environment and health. She has thirty years of teaching and professional experience, and has authored/co-authored more than two hundred technical publications, including ten books and a dozen book chapters. Prior to becoming full-time at UdelaR, she worked for 19 years as a freelance Environmental and Sanitary Engineer.

Contents

Preface

Sound is an omnipresent agent in our society. Thus, so is noise. Occupational, social (or voluntary), and environmental (or passive) exposure to noise is a part of our lives. Research on noise pollution has inspired a wide set of researchers to study this interdisciplinary subject. Sources, causes, effects on people and other animals, control, regulations are some of the multiple aspects of the problem. This book offers current approaches to some of these topics: traffic noise, the noise of the neighborhood or underwater noise; occupational noise exposure in health emergency services; new approaches and materials to model and control noise problems. Acoustic materials are widely used by noise control engineers for NVH (Noise Vibration and Harshness) reduction in all fields of transportation. Noise has various effects on comfort, performance, and human health. For this reason, noise control plays an increasingly central role in the development of modern industrial and engineering applications.

Nowadays, the noise control problem excites and attracts the attention of a great number of scientists in different disciplines. Indeed, noise control has a wide variety of applications in manufacturing, industrial operations, and consumer products. The key to success for controlling noise inside interiors in the transportation field is to study and apply innovative materials with outstanding thermo-acoustic and mechanical properties. For this purpose, the book aims to present a collection of studies and applications in the field of innovative materials for interior acoustics in transport vehicles, for researchers and acoustic specialists.

The book is organized into 8 chapters written by specialists in the acoustic sector who use numerical and experimental techniques for noise reduction in the surrounding environment also through the study of innovative materials capable of reducing the effects on humans considered harmful to health.

The book means to supply the reader, both student and researcher or teacher, with extensive knowledge of the latest research results in innovative materials for acoustics and new approaches to face such an old issue as noise pollution is. The editors are sincerely grateful to all the authors who agreed to share, with their professionalism and experience, their latest research results contributing to the success of this book.

Daniela Siano
Istituto Motori CNR,
Italy

Alice Elizabeth Gonzalez
Universidad de la Republica,
Uruguay

Section 1

Noise Pollution

Chapter 1

Influence of Noise in Ambulance Vehicles on Emergency Service Personnel

Jānis Indulis Dundurs and Inka Janna Janssen

Abstract

Every day, noise is a ubiquitous potential hazard to our body. Importance is already dedicated early in history and still continues by steady investigations in terms of protecting the personnel in loud environment. "Worldwide, 16% of hearing loss in adults is attributed to occupational noise." Noise-induced hearing loss (NIHL) is a sensorineural hearing loss, explained by permanent threshold shift of hearing sensitivity. NIHL not only affects the auditory system but also has psychosocial effects and is proved to have interference with general health by sleep disturbances or cardiovascular symptoms. This study aims to detect and define the sound pressure levels that ambulance service workers are exposed to during their shifts in ambulance vehicle, especially with the focus on differences during signal and non-signal use and different speed levels and determining whether the noise has hazardous character. The collection of study data is composed of two parts. The first part is the indication of noise level in the ambulance vehicle with the help of a sound level meter. The second part included a questionnaire that constituted 14 questions sent electronically. In total, 207 workers responded.

Keywords: Noise-induced hearing loss, emergency service personnel, ambulance vehicle

1. Introduction

Historically, the earliest descriptions about the danger of noise-induced hearing loss (NIHL) were described in 1713 by the Italian physician Bernardino Ramazzini (1633–1714). In his book "De Morbis Artificum" (Diseases of Workers), he firstly demonstrated the impact of hearing loss together in a relation to prolonged exposure to noise by his observations based on his examinations on coppersmith workers who were constantly exposed to noise and gradually suffered from hearing loss [1].

During the eighteenth century with onset of rapid industrialisation, the incidence of NIHL increased drastically and lead to the first ideas of preventive actions.

Almost 200 years later, the Hungarian biophysicist Georg von Békésy (1899–1972) analysed the travelling wave of sound in the cochlea, for which he received a Nobel Prize in 1961 and simultaneously set the cornerstone for the start of investigations of noise and hearing loss in relation to exposure time [2].

The origin of the noun “noise” is found in Latin language from the term “nausea”, which later via detours through French language was introduced as “noise” to the English language [3]. Both words have much more in common than suspected before.

Substantially, there is no difference between sound and noise. But enlightening the differences more closely, sound refers to the sense of perception that usually occurs on voluntary basis and delights the listener as it is for example by listening to music. On the other hand, noise is defined as an unwanted sound that may cause displeasure, annoyance and pain or, referring back to its word origins, nausea.

Investigations have shown that continuous noise exposure has an enormous damaging impact not only on hearing but also on the general health status of the population.

Although preventable, NIHL is one of the most widespread irreversible occupational diseases worldwide and thus was declared as a serious occupational hazard [4].

Several studies gave evidence that noise creates physical and psychological stress, commonly presented as reduced assessment, sleep disturbances, cardiovascular dysfunction and mental health alteration [5, 6].

The protection of health and safety from hazards at work should be our all interest. Therefore, our research is aimed at evaluating the impact of occupational noise on hearing, general security of health, quality of life and productivity of those working in stressful environments shown at the example of emergency service working personnel, who give constantly their best to protect and save our health during emergency.

2. Literature review

2.1 Basics of acoustics

Noise can be described as rapid fluctuations in atmospheric pressure, which affects the human body as vibrations that are perceived by the human ear and finally can be classified as sound.

Sound propagates as a pressure wave and is able to travel through any elastic medium (e.g., air, water, wood, and metal).

Important units for measurements of noise attributes are hertz (Hz) and decibels (dB), and together with some basic knowledge of physics of waves, frequency, wavelength, amplitude, refraction, absorption and transmission, we are able to understand the behaviour of noise and can develop controls and preventions. When molecules start to move due to atmospheric pressure changes, the moving air molecules pass their energy on to neighbouring molecules, which results in the spread of their energy over and over until an increasingly larger volume is created. This principle can be compared to the ripples when a stone is thrown into water. These described pressure changes are detected by the eardrum, which in return vibrates as response. In return, the vibrations are further transferred to the middle ear, which is constructed of three tiny bones facing towards the fluid-filled inner ear. The inner ear contains tiny inner and outer hair cells, which convert the vibrations into electrical nerve impulses that then are sent to the brain. Finally, the brain is then able to process these impulses into meaningful sounds [5, 7].

The perception of loudness of a sound is determined by two factors: sound pressure and frequency. The frequency (number of vibrations per second) is related to “pitch”. The higher the frequency, the shaper the sound heard by the subject [7].

Important issues about noise perception are as follows: Sound pressure levels are measured in (dB). They describe the amplitude of the sound waves. They are related to the loudness of the sound. The A-weighted sound pressure levels are measured in dB(A). A-weighting considers the non-linear response to sound of the human ear, and also its non-homogeneous response to sounds of different frequencies and intensities. This level is determined by using a standardised weighting at different frequencies and then, summing logarithmically these sound pressure levels. The A-weighted sound pressure levels better represent the auditee’s perception of noise. They are used for many applications, from community noise ordinances to occupational noise exposure regulations.

2.1.1 Noise as a dangerous hazard

Every day, we are naturally exposed to loud, distracting and possibly hazardous noise. A common experience for everyone may be the example of continues ringing after a great concert or muffled sounds after working with loud tools (chainsaw, grass cutter, etc.).

Noise at prolonged exposure at 80 dB has unsafe effects to the auditory system but also to general health [5, 7, 8].

Studies proved that the risk for NIHL increases exponentially in noise-exposed population, who are exposed to noise level beyond 85 dB(A) for a prolonged time [9].

Table 1 shows critically how noise is correlated with health that is shown in three different stages of noise levels in dB(A).

At the example of “Conversation”, it can be nicely illustrated in what manner noise level has an impact on health.

A standard conversation is measured at approximately 50 dB(A), which at a prolonged exposure may lead to mental reactions (e.g., low concentration and annoyance); at 80 dB(A), for communication, the voice needs to be elevated remarkably that interferes with health shown in physical reactions (e.g., hypertonus); and at 90 dB(A), communication is not possible anymore, which in return in long term is unbearable and triggers pain threshold.

2.1.2 Noise exposure and limit values

Determining the limit of noise exposure is crucial to take three components in consideration:

1. Worker (genetic predisposition)

2. Character of noise: sound pressure level and frequency

3. Duration of exposure

Generally, the potential and stage for hearing loss by noise are related to the workers’ duration of noise exposure and stage of noise loudness.

Halving acoustic energy can be done reducing sound pressure level by the 3 dB or halving the exposure time [10].

For better understanding, see the following examples. These noise exposures are the same:

Noise level and body reaction	Type of noise	Sound pressure levels in dB(A)	Sound sense
I 30–65 dB(A) Mental reaction	Fine ticking of a clock, whispering	30 dB(A)	Very quiet
	Library, bedroom at night	40 dB(A)	Pretty quiet
	Conversation	50 dB(A)	Normal
	Quiet office	60 dB(A)	Moderate to loud
II 65–90 dB(A) Physical reaction	Shouting, car in 10 m distance	70 dB(A)	Loud to very loud
	Street noise in heavy traffic	80 dB(A)	Very loud
III 90–120 dB(A) Hearing loss, ear pain	Loud factory hall	90 dB(A)	Very loud
	Car horns in 7 m distance	100 dB(A)	Very loud to unbearable
	Full symphony orchestra	110 dB(A)	Very loud to unbearable
	Jet engine, live rock band	120 dB(A)	Unbearable to painful
		130 dB(A)	Intolerable

Table 1.
Overview of noise level and impact on human body.

- 80 dB for 8 h
- 83 dB for 4 h
- 86 dB for 2 h
- 89 dB for 1 h
- 92 dB for 30 min

International standards recommend an “equivalent sound pressure level of 85 dB(A) at 8-h working day average as the exposure limit for occupational noise” for preservation of the personnel’s hearing when working in a noisy environment. However, in reality, it shows that this limit does not guarantee safety, especially for the hearing system of workers, since 80 dB(A) is already indicating harmful effects [11].

Therefore, Noise at Work Regulations recommend a “three action levels for occupational noise level” depending on equivalent noise level for 8-h working day (see **Table 2**).

2.2 Overview of hearing loss

Hearing loss can be categorised depending which parts of the hearing system are damaged. There are three basic types of hearing loss: conductive hearing loss, sensorineural hearing loss, and mixed hearing loss [12].

Conductive hearing loss occurs due to damage of outer structures of the auditory system.

The sound waves are not properly conducting through the outer ear canal, the eardrum and ossicles of the middle ear. It is characterised by a reduction of sound level perception or the ability to hear weak sounds. Good treatment options are surgery or medication depending on issue.

Action level	$L_{Aeq}8h$
First action level (minimum) provide protection	80 dB(A)
Second action level mandatory protection	85 dB(A)
Maximum exposure limit value	87 dB(A)

Table 2.
Three action levels for occupational noise level.

Causes of conductive hearing loss are as follows:

- Fluid in the middle ear
- Ear infection (otitis media or otitis externa)
- Poor eustachian tube function
- Trauma, e.g., perforated eardrum
- Obstacle such as cerumen, tumour or foreign body

Sensorineural hearing loss appears in case of damage to inner structures such as cochlea or to the nerve pathways from the inner ear to the brain.

Characteristically, it is described by the reduced ability to hear faint sounds. Even when speech is loud enough to hear, it may still appear to be unclear or sound muffled. Unfortunately, there is no treatment option.

Examples:

- Ototoxic medication
- Genetic or hereditary
- Ageing
- Head trauma
- Exposure to loud noise

2.2.1 Noise-induced hearing loss

NIHL is one of the most common occupational illnesses, because it is often ignored since there are no visible effects or pain sensation in early stages.

Factors that especially predispose one to NIHL are found in high-frequency noise exposure, which is known to be much more harmful than low-frequency noise, and also continuous stimuli are more damaging than interrupted stimuli.

NIHL starts with a temporary threshold shift (TTS). Physiologically explained, while the ear is exposed to the noise, there originates a release of ATP from stria cochlearis, which provides the necessary energy for the function of the hair cells. In case of prolonged elevated noise exposure, a mismatch occurs between energy supply and consumption, as the ATP needs to reach the hair cells by diffusion. The hair cells get tired due to energy depletion and result to be less sensitive. This leads to a shift in hearing threshold but has still potential to recover completely when the harmful stimulus is removed [13].

If the exposure to the harmful noise still continues, the cellular integrity of the hair cells of the Corti organ disrupts gradually and ultimately the nerve fibres that innervate the hair cells will disappear, thus resulting in permanent threshold shift (PTS), and henceforth, irreversible hearing loss at higher frequencies will be noted. In most cases, it is described affecting both ears symmetrically [14].

NIHL can be classified into four stages: mild, moderate, severe and profound [15].

1. Mild NIHL is a high-frequency hearing loss of sounds between 20 and 40 dB.

2. Moderate NIHL, between 40 and 60 dB.

3. Severe NIHL, between 60 and 80 dB.

4. Profound NIHL, greater than 80 dB.

Limitations in hearing are evident when listening to high frequencies. First noticed problems are trouble understanding speech during present background noise. NIHL progresses gradually, and people have difficulty understanding high-pitched voices (e.g., women and children) even in quiet conversational situations, whereas conversation on the telephone is generally unaffected.

TTS slowly progresses to PTS, post exposure tinnitus, and TTS serves as warning signs of impending permanent NIHL [14].

2.2.2 Audiogram

Hearing loss is detectable by performing an audiogram and is presented as a graph that shows the weakest sounds a person can hear at different frequencies. It can be used to detect the concrete severity of sensorineural hearing loss or for check-up reasons.

For TTS, there is a chance that the shift regresses again after the noise is removed.

Therefore, the sound pressure level during the recovery period is kept below 70 dB(A) and a recovery time of at least 10 h.

In case the recovery period is not respected, an accumulation of the individual TTS may occur and leads to PTS, which can be detected in the audiogram (see **Figure 1**).

NIHL is typically shown with selective loss of hearing at around 4000 Hz, which is apparent in the audiogram as a notch-like depression.

If exposure to harming noise is continued, the notch gradually deepens and widens. It can also take over to the middle frequencies. In very severe cases, even the lower frequencies may eventually become involved [5].

2.3 Consequences of NIHL

Consequences of NIHL may severely interfere with both social and occupational environment. NIHL as a reason for limited communication ability with co-workers and family may develop anxiety, irritability and decreased self-esteem resulting in loss of productivity and, eventually, social isolation.

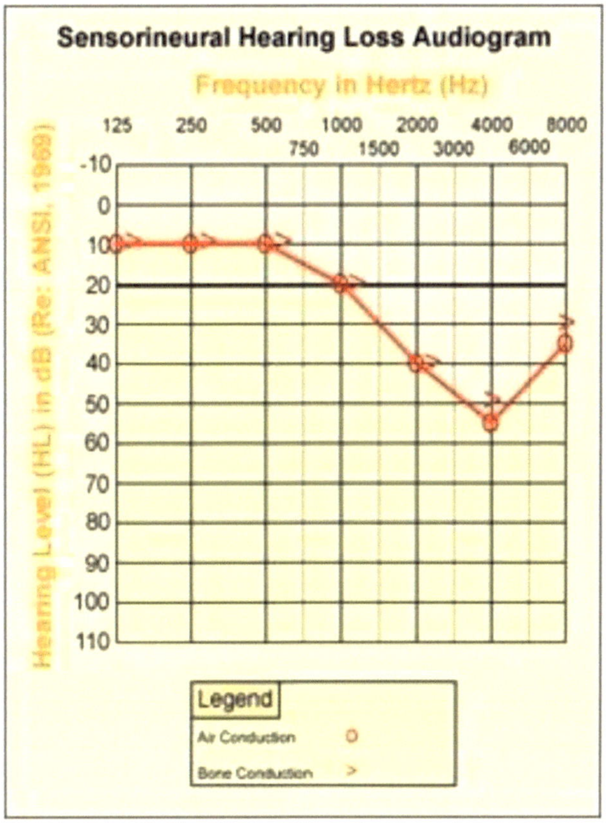

Figure 1.
Audiogram characterising early NIHL.

In terms of safety, NIHL often runs together with a reduced ability of assessment and to monitor work environment such as warning signals or equipment sounds and immensely increase the danger of injuries.

2.3.1 Health risks

Noise creates physical and psychological stress that can interfere with health leading to extra aural health risks. Common early symptoms can be found psychosocially as sleep disturbances, concentration difficulties and clumsiness.

Depending on the extent of sound pressure level, it can influence the vegetative system by a shift in favour towards the sympathetic nervous system. Examples are tachycardia, hypertonia, tachypnoea, increased adrenal secretion of stress hormones such as cortisol and decrease in gastric secretion for protection of gastric mucosa. In the long term, these symptoms have potential to interfere severely with health [16].

3. Methods and materials

3.1 Study design

The data were collected through a retrospective cohort study using a questionnaire with a total sample size of 207 workers from two main emergency ambulance service centres located in Riga, Latvia, and Aurich, Germany.

Additionally, the noise level was detected by measurements with a sound level meter.

3.2 Description of data collection

Materials of use:

1. Questionnaire (14 questions)

2. Sound level meter

The collection of data is divided into two parts.

The first part includes the collection of basic personal and working information via a questionnaire of 14 questions. Hundred and five Latvian and hundred and two German emergency service workers responded to the questionnaire. The questionnaire was shared as electronic survey at http://www.visidati.lv and also distributed as printed paper to the different working stations. The collection of personnel information started in January 2016 and ended in May 2016.

The research data were collected and statistically processed in Microsoft Office Excel 2010 and SPSS 22.0.

The second part refers to the measurement of noise level by use of a sound level meter reviewing non-signal and signal noise exposures during 12-h shifts and by this giving the basis to analyse the average noise level that an emergency service personnel is presented to.

The collection of measurement started in January 2016 and ended in February 2016.

Digital sound level meter model LUTRON SL-4013 conforms to IEC 651 Type 2 with 0'. Noise was measured using a standard microphone head that was placed in the front passenger compartment of an ambulance during emergency driving. The equipment was programmed to collect data in fast mode, using the weighting curve "A". Also, a protective foam in the microphone in order to minimise the other noise effects was used.

The measurements were recorded during 20 emergency trips with a duration range from 10 to 15 min. These measurements were performed in different days, periods and shifts. The noise levels were carefully recorded at different velocities under the following conditions:

- Asphalt street and good surface conditions of the street
- Measuring device placed in the centre of the cabin at level of the ears of workers
- Taking measured number at stable driving of 50, 70 and 100 km/h
- Mostly free field as surrounding (no high density of high houses)
- No talking, no funk communication
- Radio turned off
- Windows closed
- No rain, calm wind

The ambulance car is analysed in terms of technical specification and physical dimensions.

For proper comparison of the ambulance service in Germany and Latvia, we chose similar cars in model and age.

Both countries use the Mercedes-Benz, Sprinter 315, CDI model, 4-door, manual and manufacturing year 2010 (Riga, Latvia) and a similar model from year 2012 (Aurich, Germany) This model is a standard Sprinter with high ceilings. The front cabin design layout constituted likewise.

The sirens are located bilaterally on the roof and front spoiler of the ambulance car. The type of sirens and frequency for Latvia and Germany differ especially in frequency of sound melody. Germany is using sirens of type "Martin-Horn 2298 GM" DIN 14610 EC with a 4′ membrane-bell and sound pressure level of 125 dB(A) at a distance of 1 m.

In Latvia, there is no standardised sound melody throughout a signal trip. During signal trips, the driver can choose manually between different frequencies.

4. Research results

4.1 Sound level meter measurements

The average sound levels based on the measurements performed during numerous emergency trips are as shown in **Table 3**.

In Germany, the minimum noise level is measured at 50 km/h without signal use with 63.5 dB(A) and the maximum is measured at 100 km/h with signal use with 84.8 dB(A).

In comparison, in Latvia, the minimum is measured at 50 km/h without signal use with 67.3 dB(A) and the maximum is measured at 100 km/h with signal use with 90.7 dB(A).

For both countries, it is noticeable that the noise level during signal use is enormously elevated than during trips without signal use. Comparing the Latvian with German emergency cars, non-signal trips are measured with a higher average noise level with an average difference of 5.2 dB(A) and during signal use the noise level in Latvian emergency cars is also higher by a difference average of 2.4 dB(A). This means, the Latvian emergency personnel is exposed to an overall higher noise level during emergency trips than German emergency personnel.

4.2 Questionnaire

The research included in total count 207 emergency workers from different emergency service centres. Hundred and two German and hundred and five

	Germany		**Latvia**	
	Without signal	**With signal**	**Without signal**	**With signal**
Average	66.5	84.7	71.9	86.6
50 km/h	63.5	84.4	67.2	83.3
70 km/h	65.2	84.8	72.4	85.9
100 km/h	71.4	84.8	76.1	90.7

Table 3.
Measured sound pressure level in dB(A) according to speed and considering all data together, for Germany and Latvia.

Latvian emergency workers answered fourteen questions in an electronic form at http://www.visidati.lv or in printed version.

4.2.1 Population characteristics

In total, the respondents are defined by 116 (56%) men and 91 (44%) women aged between 18 and 65 years.

In Germany, the majority of personnel is formed by men (35%), while in Latvia, the majority is dominated by female workers (30%) (see **Figure 2**).

The age distribution shows that the Latvian emergency personnel in general are composed of a rather young team in the age range of 18–30 years (18–25 years = 57.1%, 26–30 years = 30.5%), and in return, the German personnel show a wider range of age distribution, which majorly is observed to be between 18 and 40 years (18–25 years = 46.1%, 26–30 years = 22.5%, 31–40 years = 16.7%). Consequently, the emergency workers in Germany are older compared to the Latvian emergency workers (see **Figure 3**).

Relating the personnel's age and years of employment, a correlation is apparent. The vast majority of the young Latvian workers have been working for 1–5 years (84.8%) and then an abrupt decrease of employment time by approximately 80% is seen, while the investigated German ambulance service also has its peak employment time at 1–5 years (54.9%) but then gradually decreases by 50% (see **Figure 4**).

According to the amount of shifts per week and the density of emergency occurrence, both countries have four shifts of 12 h during a 7-day working week and parallels are seen for the average amount of trips that are set at approximately six non-signal and five signal trips for both countries.

Evaluating the amount of hours the worker is exposed to noise during trips, the German ambulance service personnel are approximately 1.5 h (63 min) longer exposed to noises from signal and 0.71 h (43 min) longer exposed to non-signal noises during a 12-h shift. Comparing both countries for their total emergency

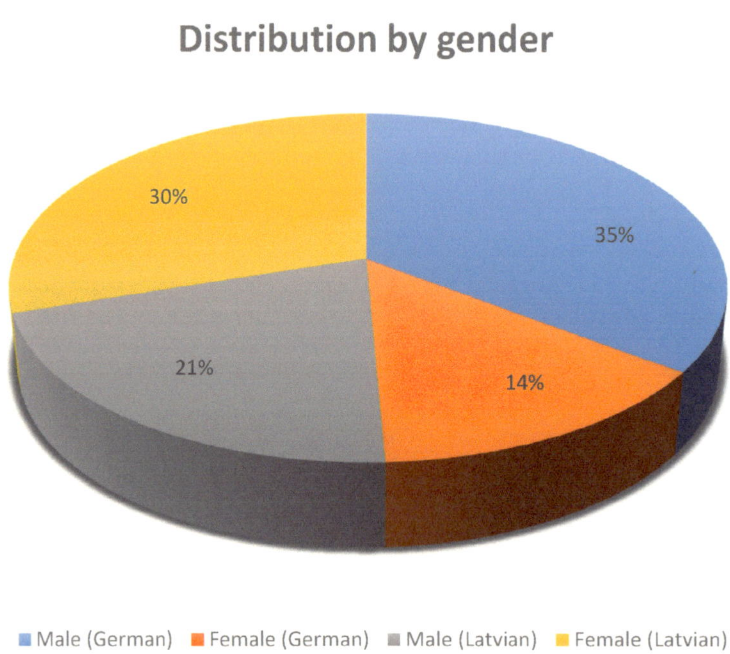

Figure 2.
The percentage of survey population by gender.

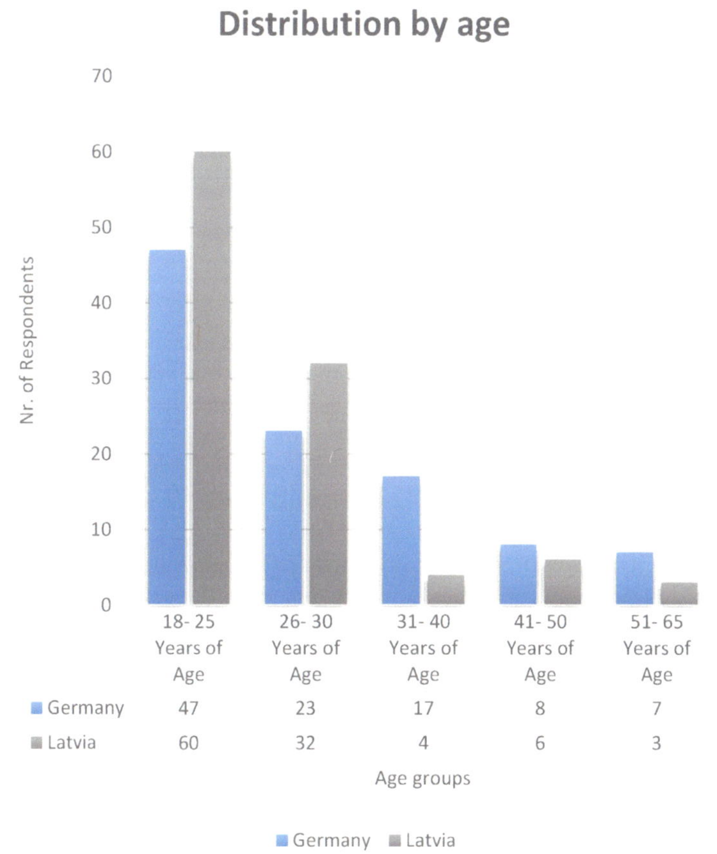

Figure 3.
The distribution in absolute numbers of survey population by age, Latvia and Germany in comparison.

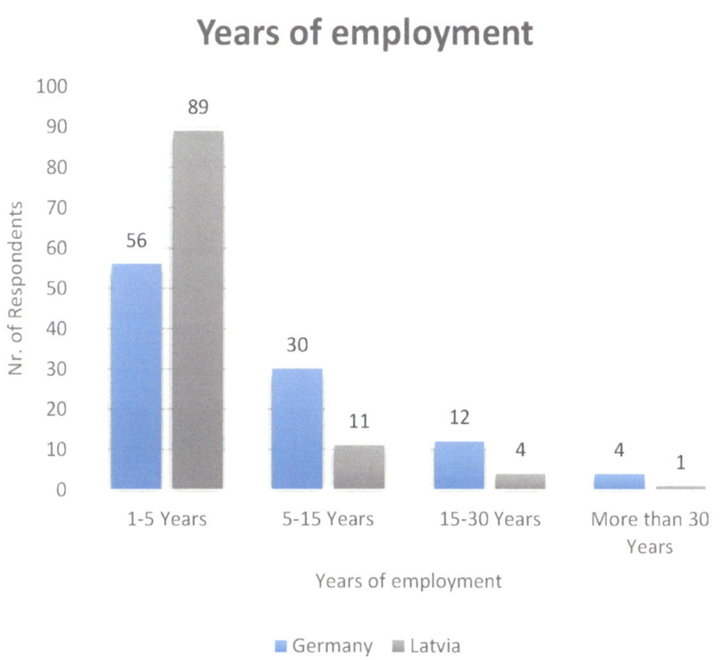

Figure 4.
The distribution in absolute numbers of survey population by years of employment, Latvia and Germany in comparison.

trip-related noise exposure during a 12-h shift, German personnel are in total 68% exposed to noise and Latvian personnel in total 53% (see **Figure 5**).

The time that the worker is not sitting in the car and presented to the evaluated noise is categorised as "other" in **Figure 5**, which stands for the time, e.g., in the hospital, patient house or guardhouse. It is impossible to measure these noise levels; nonetheless, it should be taken into consideration, since presentation to noise is ubiquitous and affecting the body.

4.2.2 Symptom prevalence

In the questionnaire, respondents were allowed to choose more than one symptom and indeed, most respondents indicated more than one symptom. In comparison, Germans assigned 1–2 fitting symptoms and Latvians choose 2–3 (see **Figure 6**).

For both countries, a common pattern of complaints and also highest incidences were found in following two symptoms: difficulties of understanding during background noises (Germany 30.2%, *N* = 42 and Latvia 26.6%, *N* = 46) and tinnitus (Germany 38.8%, *N* = 52 and Latvia 23.8%, *N* = 41).

Other similarities but with lower frequency are given for hyperacusis (Germany 6%, *N* = 6 and Latvia 5.8%, *N* = 10), pain or pressure in the ear (Germany 13.4%, *N* = 18 and Latvia 13.4%, *N* = 23) and difficulties understanding speech, particularly women and children (Germany 7.9%, *N* = 11 and Latvia 10.4%, *N* = 18).

Comparing main differences, Latvian emergency workers show a much higher incidence of symptoms such as vertigo (23.8%, *N* = 41), changes in sound perception (7.5%, *N* = 13), difficulties in determination of sound direction (5.8%, *N* = 10) and difficulties using the phone due to poor understanding of the partner (13.3%, *N* = 23).

German emergency personnel showed higher prevalence only for difficulties understanding electronic audio devices such as TV and radio and thus the need to increase the volume (20.9%, *N* = 28).

Concluding, Latvian emergency personnel clearly dominate in 8 from 10 auditory symptoms with higher absolute number.

For a closer accurate evaluation risks for NIHL, also an average noise exposure during free time was requested giving a defined range from 1 (low noise exposure) to 10 (high noise exposure). Both indicated an average free time exposure to noise at 5.

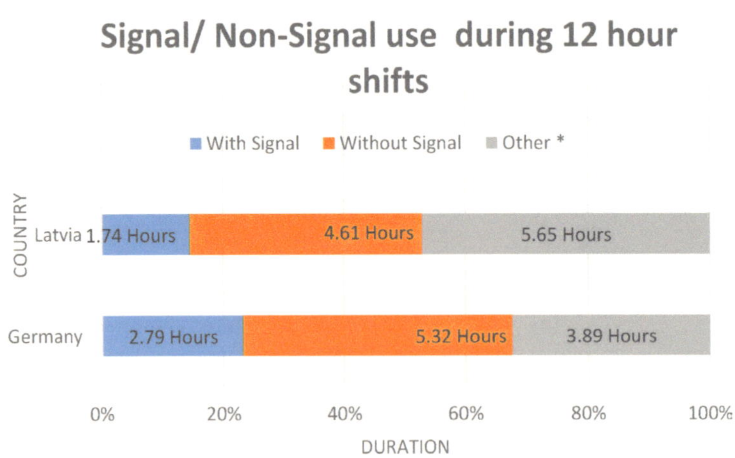

Figure 5.
*Time of exposure of the survey population to signal and non-signal trips during a 12-h shift. *Others include the time during the 12-h shift outside the emergency car.*

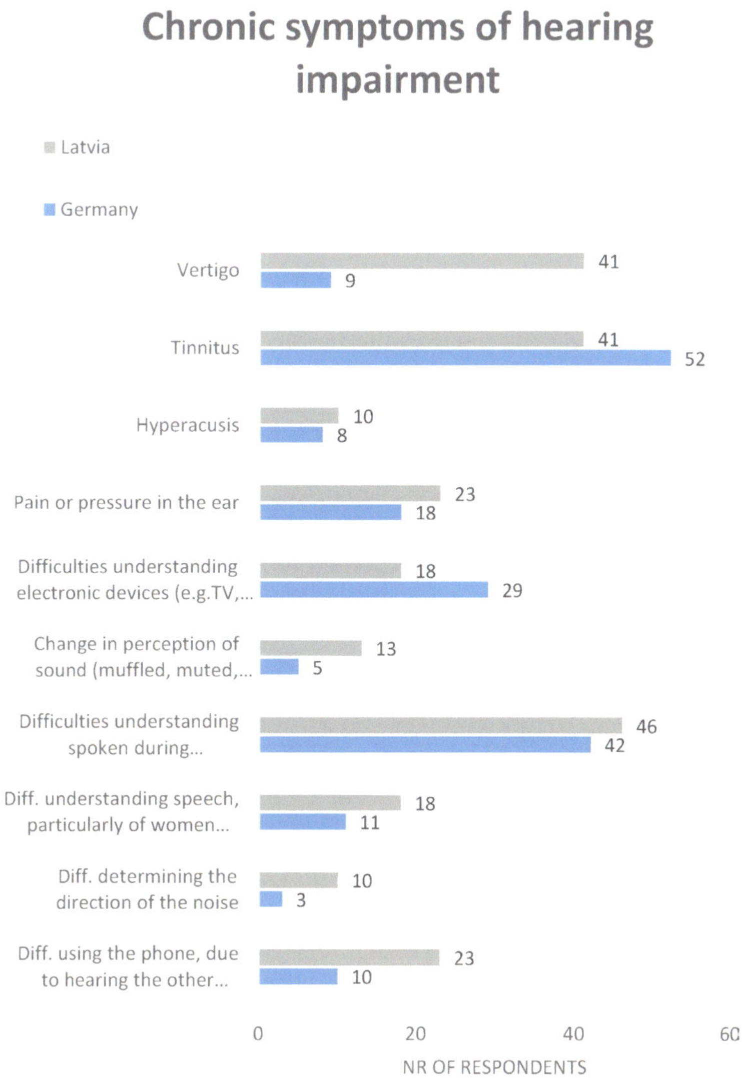

Figure 6.
Prevalence of chronic hearing symptoms for each country of the survey population in absolute number, Germany and Latvia in comparison.

4.2.3 Statistical data

Statistical investigations for relations according to the study showed that there is no significant relation between countries, age, gender or length of employment towards symptoms ($p > 0.05$) and thus cannot be attributed to the general population.

5. Discussion

When persons with normal hearing are exposed to high noise levels over a prolonged period of time and by this reaching or exceeding the limit of permissible noise level exposure equivalent of 85 dB(A) during 8 h, a shift of hearing threshold may result. Under a threshold shift is meant an average deterioration of hearing of 10 dB(A) or more in the frequency ranges of 2000, 3000 and 4000 Hz in both ears, defined by Occupational Safety and Health Act (OSHA).

This deterioration of hearing can be of a temporary nature (TTS), or in opposite at continuous exposition can result to a permanent threshold shift (PTS) and hearing loss.

The amount of hearing loss results from the sound pressure level, the duration of exposure, the frequency of noise and the individual predispositions.

The study focuses on the distribution of symptoms that determine the current state and assess the future trend of NIHL risks. The Latvian ambulance service personnel are exposed up to 5.2 dB(A) (non-signal) and 2.4 dB(A) (with signal) louder noise than German personnel. Both countries demonstrate an exposure to hazardous noise level of approximately 85–90 dB(A) during signal trips, which reaches and partly exceeds the exposure limits of 85 dB(A).

Sound measurements of this study show that during non-signal trips, the noise pressure level varies depending on speed by 2–6 dB(A). The faster the speed level, the greater the noise level. During signal trips, for Latvians, the increase in noise level is by 2–4 dB(A) depending on speed level seen, but for Germans, the noise level stays almost constant at different speed levels.

However, during a 12-h shift, the Latvian survey population is exposed for approximately 2 h to signal trips with an average noise level of approximately 87 dB(A) and the German survey population approximately 3 h to signal trips with an average noise level of 85 dB(A).

Referring to OSHA regulations, both countries are not exceeding the limit of permissible noise level exposure equivalent. Thus, the exposure to noise during emergency trips with signal is considered to be safe for the auditory system.

Nonetheless, especially the Latvian emergency personnel indicate a great dominance for auditory changes, as clearly shown in my study data.

Possible explanations for the contrary facts may be found when considering the sirens of the ambulance vehicles. The frequencies of ringtones that can be selected in Latvian cars are usually higher and thus more harmful to the hearing system. Also the majority of streets are in rather poor condition, which increases the noise level by its vibrations. Furthermore, accumulations of numerous unrecovered TTS by short and extreme fluctuations of noise level may also trigger NIHL. Moreover, the natural limitations of the study need to be taken into account. Firstly non-job-related noise exposure such as listening to a walkman loudly for long time or being a member of an orchestra and giving a concert has a great impact on hearing, which limits the accuracy of the study. Secondly, during the last 5 years, both ambulance services invested enormously into new cars and equipment. Thus, the symptoms can be a result from the older cars, where presumably the noise level must have been presented far louder.

5.1 Preventive measures

Personal:

- Regular medical examinations of workers
- Personal protective devices (e.g., filter-type earplugs)
- Education of both workers and the management staff in order to prevent NIHL
- Planning and organisation to avoid streets of bad quality, which produce excessive noise or need of prolonged signal use due to a crowded traffic

- Intelligent planning of the duty roster to provide rest from loud noise exposure
- Keeping the noise level and its exposure during leisure time safe
- Audiogram check-ups to make the personnel more aware of the auditory status

Vehicle:

- Acoustic insulation and sound proofing to doors, walls and ceilings
- Fixing all loose equipment in the cabin for safety reasons but also noise reduction
- Positioning of sirens as far away as possible from the personnel, e.g., front of the spoiler

6. Conclusion

1. NIHL is one of the oldest and most common occupationally induced health issues worldwide.
2. Common pattern and highest prevalence for auditory symptoms for both Latvian and German ambulance services are:
    - Difficulties of understanding during background noises
    - Tinnitus
    - Vertigo
    - Difficulties understanding electronic audio devices such as TV and radio and thus the need to increase the volume
3. The Latvian ambulance service personnel have a higher risk of developing NIHL reasoned by high frequency of sound melody of the sirens and exposure to higher sound level during signal trips, caused by poor street conditions.
4. For both countries, the noise level is remarkably elevated during signal trips compared to non-signal trips.
5. Speed level influences the noise level during trips without signal by 2–6 dB(A). The higher the speed, the higher the noise level during non-signal trips.
6. During emergency trips with signal use, the noise level is reaching and partly exceeding the safety limits of 80–85 dB(A).
7. Education of the ambulance workers and management about preventive measures, the importance of NIHL development and risk as well as regular audiometry check-ups are needed.

Author details

Jānis Indulis Dundurs* and Inka Janna Janssen
Department of Occupational and Environmental Medicine, Riga Stradiņš University, Latvia

*Address all correspondence to: janis.dundurs@rsu.lv

References

[1] Rabinowitz PM. The Public Health Significance of Noise-Induced Hearing Loss. Berlin, Germany: Research Gate. 2012:13-25

[2] T.V.A. Safety Institute of Australia Ltd. OHS Body of Knowledge, Physical Hazards: Noise [Online]. 2012. Available from: http://ro.uow.edu.au/cgi/viewcontent.cgi?article=3381&context=hbspapers [Accessed: 23 September 2016]

[3] Word Central, Student Dictionary. Merriam-Webster [Online]. 2007. Available from: http://www.wordcentral.com/cgi-bin/student?noise [Accessed: 06 October 2016]

[4] Prevention of noise-induced hearing loss. Report of a WHO-PDH Informal Consultation. Geneva: WHO. 28-30 October; 1997

[5] Dennis P, Driscoll P. OSHA Occupational Safety and Health Administration [Online]. 2011. Available from: https://www.osha.gov/dts/osta/otm/new_noise/index.html#hearingloss [Accessed: 21 September 2016]

[6] Jansen G. Zur Nervösen Belastung Durch Lärm. [To the nervous load due to noise] Darmstadt: Steinkopff Verlag; 1967

[7] Bundesministerium für Umwelt, Naturschutz, Bau und Reaktorsicherheit [Online]. 2016. Available from: http://www.bmub.bund.de/themen/luft-laerm-verkehr/laermschutz/laermschutz-im-ueberblick/was-ist-laerm [Accessed: 22 September 2016]

[8] National Institute on Deafness and Other Communication Disordern (NIDCD) [Online]. 2015. Available from: https://www.nidcd.nih.gov/health/noise-induced-hearing-loss [Accessed: 21 September 2016]

[9] Boateng CA, Amedofu GK. Industrial noise pollution and its effects on the hearing capabilities of workers: A study. African Journal of Health Sciences. 2004;**11**(1-2):55-60

[10] Health and Safety Executive. Controlling noise at work. In: The Control of Noise at Work Regulations 2005. Guidance on Regulations. Richmond, Surrey: HSE; 2005

[11] Konrad Rydzynski DTJ. Directorate-General for Health & Consumers [Online]. 2008. Available from: http://ec.europa.eu/health/ph_risk/committees/04_scenihr/docs/scenihr_o_018.pdf [Accessed: 24 September 2016]

[12] Kasper FH. Harrison's Principles of Internal Medicine. New York: McGraw-Hill Professional; 2015

[13] Muñoz DJ, Kendrick IS, Rassam M, Thorne PR. Vesicular storage of adenosine triphosphate in the guinea-pig cochlear lateral wall and concentrations of ATP in the endolymph during sound exposure and hypoxia. Acta Otolaryngology. 2001;**121**(1):10-15

[14] Mathur NN. Medscpae [Online]. 2016. Available from: http://emedicine.medscape.com/article/857813-clinical#showall [Accessed: 24 September 2016]

[15] Marisol Concha-Barrientos DC-LKS. World Health Organization [Online]. 2004. Available from: http://www.who.int/quantifying_ehimpacts/publications/en/ebd9.pdf [Accessed: 21 September 2016]

[16] Heckl MMHA. Taschenbuch der technischen Akustik. 2nd ed. New York: Springer; 1994

Chapter 2

Neighborhood Noise

Jiho Lee

Abstract

Listening to sounds in everyday life is an important factor in a human life. You can talk, listen to music, and enjoy nature through sound. However, due to adverse effect, listening to unwanted sounds continuously may cause noise-induced health disorders. Noise is an inevitable pollution factor in modern society, and its severity is increasing day by day. In addition, as the improvement of the economic level and the demand for the calm quality of life are increasing, the noise-related problem is emerging as a continuous social issue. Most of the problems associated with noise are mental, especially in developed countries, where social issues with the neighborhood noise are getting worse. The severity of noise-related problems is associated with the characteristics of noise, personal sensitivities, and vulnerable groups, but continued exposure can adversely affect not only health but also sociocultural, ethical, and economical aspects. However, the knowledge of the direct and indirect effects of noise pollution on health is still insufficient. Due to these limitations, it is difficult to establish reasonable standards for resolution and therefore requires more scientific research works.

Keywords: neighborhood noise, annoyance, health effects, environmental burden, mediation, social issue

1. Noise pollution becomes a main problem

With the improvement of living standards, urbanization, and industrialization, noise pollution has become an environmental factor that is most frequently encountered by anyone, anytime, and anywhere in everyday life. Unlike other environmental problems, noise pollution tends to increase continuously, and the sufferings of the residents exposed to noise also increase gradually. In particular, in a rapidly developing society, poor buildings' quality, poor urban planning, and traffic noises generate more exposed to noise pollution.

Korea is successfully industrialized country. With the industrialization, noise complaints began to emerge. Noise and vibration make up 90% of the environmental disputes. Most metropolitan residents in Korea are suffering from noise pollution. And 88% of metropolitan residents expected that noise level would get worse.

The data of the nationwide environmental noise through automatic measurement network in 2018, which included major cities in Korea, were as follows (**Figure 1**). The distribution of noise level was 84.5% in the case of over 55 dBA at night (23–7 hours) and 99.9% in the daytime (8–22 hours), and some cases exceeded 75 dBA (0.95% at night, 4.54% during the day). The national average noise level was 64.6 dBA (54.6–69.9 dBA) during the night and 69.6 dBA (55.1–74.3 dBA) during the day, 5 dBA higher than during the night. Most of results exceeded the domestic standard for residential areas, 50 dBA at night and 55 dBA during the day. Because such noise level is a result of outdoor measurement, the indoor noise level might be

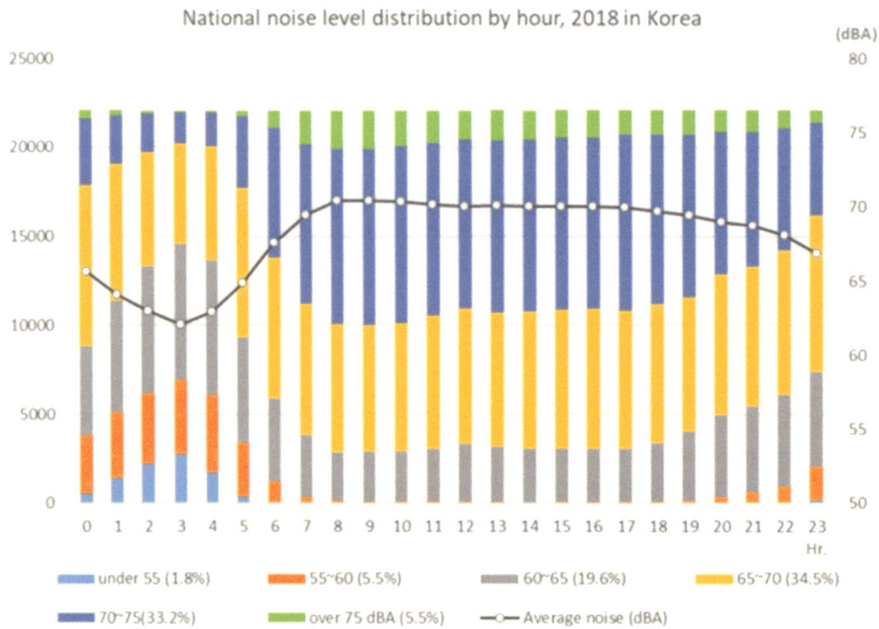

Figure 1.
Distribution and mean noise levels of nationwide automatic measurement system in 2018, in Korea: Data from http://www.noiseinfo.or.kr.

10–15 dBA lower than outdoor level usually [1]. The nationwide environmental noise level and noise-related problems in Korea are not getting better than before.

According to the International Program on Chemical Safety [2], an adverse effect of noise is defined as a change in the morphology and physiology of organism that results in an impairment of functional capacity, or an impairment of capacity to compensate for additional stress, or increases the susceptibility of organism to harmful effects of other environmental influences. This definition includes any temporary or long-term decrement of physical, psychological, or social function of humans or human organs.

Environmental noise exposure is responsible for range of health effects, including increased risk of ischemic heart disease as well as sleep disturbance, cognitive impairment among children, annoyance, stress-related mental health risks, and tinnitus. Taken together, these risks in high-income European countries account for a loss of 1–1.6 million disability adjusted life years (DALYs) – a standardized measure of healthy years of life lost to illness, disability, or early death.

The health effects of noise depend on its complexity such as time variation, frequency content, loudness, ambient noise level, type of noise, and individual difference. The lack of sufficient knowledge about the direct and indirect effects of these noises on health is limiting the provision of reasonable regulatory standards for living noise.

2. Neighborhood noise is increasing

A rapid increase in population in the city and urbanization in 1960s and 1970s prompted the need for residential construction. To cope with the demand, high-rise flats were built. Regulation on building at that time did not include the test on sound insulation in residential building, so standards were often not adequate to protect people from everyday sounds from their neighbors made [3]. Although there were some differences by country, neighborhood noise issues were mostly published after the 1980s. They revealed that economic growth and urbanization affected the neighborhood noise issues.

According to the special report of New York City in 2005, there were 410,000 noise complaints to 311. New Yorkers perceived much more from neighborhood noise and also suffered more behavioral and emotional consequences, such as difficulty sleeping and relaxing and feeling annoyed, angry, or upset compared to the nationwide population. New Yorkers were especially bothered by neighbor noises such as inadequate floor covering and slamming of doors. Of those, young children running around excessively were noise complaints that were best handled by clauses in apartment leases. These findings demonstrated that New Yorkers could not find the requisite peace and quiet in their homes that they deserve [4].

The European quality-of-life surveys were carried out examining both the objective circumstances of lives of European citizens and how they felt about those circumstances and their lives in general. The last (fourth) survey in 2016–2017 involved nearly 37,000 citizens, and respondents were asked whether they had major, moderate, or no problems with noise from the immediate neighborhood of their home. Almost one third (32%) of them reported problems with noise (ranging from 14 to 51% in individual countries), mainly in cities or city suburbs (49%) [5]. The neighborhood noise problem accounted for a large proportion of complaints related to noise and its proportion increased despite the government efforts such as campaign and legislation.

Neighborhood noise may stem from various potential sources of noise (such as ventilation systems; church bells; animals; neighbors; commercial, recreational and occupational activities; or shooting/military). As the sources might be located in close proximity to where people live, they could cause considerable annoyance even at low levels.

The main background factors of noise issues include overcrowding, developing urbanization, sprawling development, building of apartments and houses with inadequate sound insulation, increased use of electric equipment at home, increased number of recreation facilities, and lack of communication among neighbors. In addition, the calmness of a residential area depends on the noises outside the house. These main noise issues simply divided neighborhood noise into three categories: (1) noise produced by using loudspeakers, (2) noise produced during the nighttime operation of commercial facilities, and (3) daily life noise [6].

According to the report of Right to Peace and Quiet Campaign (RPQC) in 1994, at least five people a year died from noise-related conflicts between neighbors in the UK [3]. Also, 18 people had serious social problems in 2010–2020 such as arson and murder followed by conflicts related to neighborhood noise in Korea.

People could feel more annoyed if they believe the noise might harm our health or put us in danger. They could be particularly disturbed when their neighborhoods suddenly become noisy. When noises become really disturbing, it could dominate every aspects of our lives. The desire to get rid of the offending noises by almost any means possible could be overwhelming. Murder or suicide is just the end point of that process. Although only a small number of people resort to suicide or murder, many lives could get altered forever by noise problems [3].

3. Neighborhood noise problem and the related efforts in Korea

According to mediation center report, of the 137,813 telephone consultations (2012–2018), there were severe conflicts among neighbors, and 39,950 cases

(29.0%) were requested for onsite diagnosis and measurement. The mediation service demand has increased by 3.2 times from 8795 cases in 2012 to 28,231 cases in 2018. 12493 cases required on-site diagnosis and measurement. Even though the construction year varied among those cases, the slab thickness of the apartments estimated to be less than 120 mm. Of the 1271 noise measurements, 1177 (92.6%) were within the standard, and only 94 (7.4%) exceeded the regulatory standard in 2018. Of the number of onsite diagnoses and measurements received, the floor impact sound distribution was 82.8%, and in particular, "children's running or footsteps" accounted for 70.6%, followed by hammering, furniture pulling, door closing, vibrating machines, and exercise equipment. The most common air transmitted noise was generated by household appliances, followed by musical instruments, argument, pets, toilet drains, and air conditioner outdoor units.

Most of the damages reported to the Mediation Center were sleep disturbance, followed by rest disturbance, excessive protest from the victim, emotional anxiety, and learning disturbance. In case of the conflict period between neighborhoods, less than 6 months was the most frequent, and it tended to decrease over time until 2 years but increased after that.

These results showed that the victim initially responds to the neighborhood noise sensitively due to unfamiliar state of the living environment, but eventually the pattern of response improved due to changes in behavior attitude, improvement of mutual relations, and habitualization of noise. However, it is estimated that if the period gets prolonged, the damage is re-recognized when the subjective tolerable limit is exceeded (**Table 1**).

If the noise exposure persists over an extended period of time, increasing evidence suggests that more severe health consequences, such as cardiovascular diseases, may emerge as a result of prolonged physiological stress [7, 8].

Korea's standards of Environment Noise were first established in 1964 as "Pollution Prevention Act" and have gone through several revisions in the following order, Environmental Protection Act (1978), Noise and Vibration Control Act (1991), and still the revision is ongoing. The intent of the law is to preserve proper environment, which requires the establishment of various measures, such as setting

Type of impact	**Conflict period between neighborhood (years)**							
	Total (%)	**>0.5**	**0.5–1**	**1–1.5**	**1.5–2**	**2–3**	**3<**	**Others**
Total (%)	4684 (100.0)	1281 (27.3)	1114 (23.8)	809 (17.3)	269 (5.7)	452 (9.6)	538 (11.5)	223 (4.8)
Sleep disturbance	2865 (61.2)	838	705	489	163	275	333	64
Rest disturbance	662 (14.1)	160	109	118	35	63	90	87
Excessive protest	590 (12.6)	182	136	109	32	56	48	27
Emotional anxiety	315 (6.7)	55	98	56	20	36	41	9
Learning disturbance	94 (2.0)	16	28	22	4	9	13	2
Others	159 (3.4)	30	39	15	15	13	13	34

Table 1.
The receipt situation of mediation center for neighborhood noise by conflict period and type of impact in Korea (2017–2018).

environmental standards, designating an area requiring countermeasures against noise, and setting rational permissible emission standards necessary to protect the health, property, and pleasant natural environment of the people. The law determines the regulation area of living noise where control standards are needed to impose adjustment of working hours, suspension of noise producing activities, and installation of soundproofing facilities. In addition, for those who fail to fulfill the act, it enables to prohibit the use or closure of the industry. In 2010, the revised enforcement rules have stipulated the range of noise generated by human activities (**Table 2**).

As the problem of neighboring noise became more serious, the government prepared comprehensive plans to reduce living noise in 2010. The related contents are in the following paragraphs.

First, strengthening the precautious prevention: provision of regulations for surrounding noise sources for quiet facilities (schools, libraries, hospitals, elderly facilities, childcare facilities, apartment houses, etc.), recognition of the amount of the fine caused by the noise and vibration dispute; second, management of new noise sources and living noise: present management standards for the floor impact noise and noise rating system for home appliances, preparing low frequency noise management guidelines; and third, traffic noise management: expanding the supply of low-noise cars and low-noise pavement and designating traffic noise management areas.

As a result of these efforts by government departments, the standard for neighborhood noise was more strengthened than the first one. The following shows the standards for interlayer noise implemented since 2014 (**Table 3**). The inter-floor noise-related policies of other countries are centered on lightweight impact noise, and the recommendation to the perpetrator (the UK) and fine imposition (the USA and Germany) is the main method. The allowable range varies from 65 dB in Spain

Target areas	**Noise source**		**Morning (05–07) Evening (18–22)**	**Day (07–18)**	**Night (22–05)**
Living area*	Loudspeaker**	Outdoor	60	65	60
		Transmitted to indoor	50	55	45
	Factory		50	55	45
	Industry	Same building†	45	50	40
		Others	50	55	45
	Construction		60	65	50
Other area	Loudspeaker	Outdoor	65	70	60
		Transmitted to indoor	60	65	55
	Factory		60	65	55
	Industry	Same building	50	55	45
		Others	60	65	55
	Construction		65	70	50

**Area straightly within 50 m from boundary of a general hospital under the Medical Act, schools under the Elementary and Secondary Education Act and the Higher Education Act, and public libraries under the Library and Reading Promotion Act.*
***The loudspeaker installed outdoors should be used within 3 minutes at once with at least 15-minute interval.*
†The term "Same building" refers to a building in accordance with Article 2 of the Building Act, which has a roof, pillar, or wall as a whole.

Table 2.
Noise and vibration control act and related standards (dBA SPL).

Classification of neighborhood noise		Standard for neighborhood noise	
Parameter	Measuring unit (dBA)	Daytime (6 am to 22 pm)	Night (22 pm to 6 am)
Direct impact noise	Equivalent noise level (L_{eq}) for 1 minute*	43	38
	Maximum noise level (L_{max})**	57	52
Air transmission noise	Equivalent noise level (L_{eq}) for 5 minutes*	45	40

**The equivalent noise level (L_{eq}) for 1 minute and the equivalent noise level (Leq) for 5 minutes are the highest values measured in accordance with Note 3.*
***The maximum noise level (L_{max}) is considered to have exceeded the standard if the value exceeded three times per hour.*

Table 3.
Supplementary standards for neighborhood noise (Note 3).

First-step service	1. Telephone counseling	National noise information system and nationwide call center After determining the cause of neighborhood noise, conflict resolution and mitigation measures are presented
	2. Confirm respondent onsite diagnosis and related measurement acceptance	Accept: Implementation of second-step service Refuse: Noise reduction, conflict mitigation measures postal notice (end)
Second-step service	1. Additional telephone counseling	In-depth consulting on both sides If satisfied or self-solve the problem then finish
	2. Onsite diagnosis	Consultation of mitigation measures after understanding housing structure, causes of conflict, and degree of conflict Write the consultation report
	3. Onsite measurement of noise	Measurement of noise level with equipment 24 hours Completion of the consultation report

Table 4.
Work processing flows of neighborhood noise mediation center.

to 53 dB in Finland. In Korea, the law related to interlayer noise, which is stricter than that of the other countries, was enacted considering heavyweight shock, lightweight impact noise, and maximum time-weighted noise level (L_{max}).

According to a 2013 National Human Rights Commission survey, 88% was stressed by interlayer noise. The response choices included were patience (46%), request to visit (25%), report to the guard (19%), protest after visit (7%), and report to police and neighboring centers (1%).

The Neighborhood Mediation Center opened in March 2012 to prevent neighborhood noise from social issue of multiunit apartments and to settle disputes reasonably (**Table 4**).

4. Environmental noise indicators of the health impacts

The suitable indicators for policy making on the basis of the most frequently used average noise indicators in Europe are L_{den} and L_{night}. These are used widely for exposure assessment in health effect studies and noise impact assessments.

The L_{den} indicator is day-evening-night-weighted sound pressure level as defined in Section 3.6.4 of ISO 1996-1:2016. It is calculated by the A-weighted average sound pressure level measured over a 24-hour period, with a 10 dB penalty, a 5 dB, and no penalty, each added to the average level at night, evening, and the daytime period, respectively [9]. The penalties considered people's extra sensitivity to noise during the evening and night. The L_{night} indicator is the equivalent continuous sound pressure level when the reference time interval sets during the night.

In general, environmental noise is composed of complexed component such as impact sounds and impulse sounds, which make the L_{den} or L_{night} indicators hard to represent a particular noise effect. For single-event noise indicators, the maximum sound pressure level ($L_{A,max}$) and its frequency distribution can be more appropriate in specific situations, such as in the context of night-time railway, aircraft noise events, and neighborhood noise that can clearly elicit awakenings and other physiological reactions that can be determined by $L_{A,max}$. The $L_{A,max}$ indicator is maximum time-weighted and A-weighted sound pressure level within a stated time interval starting at t_1 and ending at t_2, expressed in dB [10]. Nevertheless, the assessment of the relationship between different types of single-event noise indicators and long-term health outcomes at the population level remains tentative.

All noise exposure prediction models used today estimate free-field exposure levels outdoors, and most noise abatement regulations refer to outdoor levels as well. Nevertheless, in certain cases, it would be helpful to estimate indoor levels based on outdoor values. The differences between indoor and outdoor levels are usually estimated at around 10 dB for open, 15 dB for tilted or half-open, and about 25 dB for closed windows [1].

Regarding the night noise impacts on health, below the level of 30 dB L_{night}, no effects on sleep are observed except for a slight increase in the frequency of body movements during sleep due to the night noise. There is no sufficient evidence that the biological effects observed at the level below 40 dB L_{night} are harmful to health. However, adverse health effects are observed at the level above 40 dB L_{night}, such as self-reported sleep disturbance, environmental insomnia, and increased use of somnifacient drugs and sedatives. Therefore, 40 dB L_{night} is equivalent to the lowest observed adverse effect level (LOAEL) for night noise. Above 55 dB, the cardiovascular effects become the major public health concern, which are likely to be less dependent on the characteristics of the noise. Closer examination on the precise impact will be necessary in the range between 30 and 55 dB because most will depend on the detailed circumstances of each case. The causal link between immediate physiological reactions and long-term adverse health effects is complex and difficult to prove [11, 12].

Since most of the problems of neighborhood noise are generated during the evening or at night, it is reasonable to estimate physical health effects using the noise indicators presented above. However, the actual field noise measurement results showed a few cases that exceeded the regulatory standards, and collisions among neighbors occurred even at relatively low noise levels. This means that the problems related to neighborhood noise are largely responsible for mental health effects such as annoyance and sleep disorders. It also suggests that the effects of "effect modifiers," such as differences in noise levels from the ambient noise, socioeconomic status, and personal susceptibility to noise, should be considered.

5. Health outcomes of the noise exposure

Exposure to noise can lead to auditory and nonauditory effects on health. Through direct injury to the auditory system, noise exerts auditory effects such as hearing loss and tinnitus. Noise is also a nonspecific stressor that has been shown to have an adverse effect on human health, especially following long-term exposure. These effects are the result of psychological and physiological distress, as well as disturbing homeostasis of an organism and increasing allostatic load [13].

The most common noise-related health effect is annoyance. Noise annoyance is caused by noise-related disturbances of the individual's speech communication, concentration, and performance of tasks, and it is commonly associated with negative emotional reactions, such as feelings of displeasure, anger, and disappointment. Furthermore, annoyance may give rise to physiological symptoms, including tiredness, stomachache, and stress symptoms. In fact, noise annoyance is a symptom of stress building up inside as a consequence of signals transmitted from the auditory system to the nervous system, stimulating several subsequent reactions in our bodies [14].

Since endocrine changes manifesting physiological disorders come first in the chain of cause effect for perceived noise stress, noise effects on stress hormones may therefore be detected in populations after relatively short periods of noise exposure. This makes stress hormones a useful stress indicator but in terms of the risk assessment, usually the quantitative interpretation of endocrine noise effects is often a quantitative one rather than quantitative one. The most well-known mechanism mediating the response to stress is the hypothalamic-pituitary-adrenal (HPA) axis. When the HPA axis receives a signal of a stress response, corticotropin-releasing factor is secreted from the hypothalamus, which in response releases adrenocorticotropic hormone from the pituitary gland. Adrenocorticotropic hormone then promotes the secretion of cortisol from the adrenal cortex through the blood, which triggers responses to various kinds of stress. The secretion of cortisol in response to stress inhibits the function of the HPA axis to disrupt the secretion of neurohormones and neurotransmitters as well as influencing the endocrine system, thereby disturbing homeostasis of the body, which can induce the development of various stress-related diseases [15]. Recent studies have also reported that sleep quality and noise sensitivity are not related to vascular function, but rather that night noise increases the risk of cardiovascular disease due to the increased blood pressure [7].

The associations between noise and health could be modified by several factors (effect modifiers), so individuals may therefore be more or less affected by the noise. These so-called "effect modifiers" can be demographic factors, for instance, age, sex, and socioeconomic position; personal or attitudinal factors, such as noise sensitivity and fear of the noise source; or related to the individual lifestyle and occupation, including physical activity, psychosocial health, and job strain. In addition, coping mechanisms, such as use of ear plugs or window opening behavior, and situational factors, including time of day and type of activity, may modify the effect of exposure (**Figure 2**) [16]. Identification of risk groups, individuals who are particularly vulnerable to noise, is important for assessments of public health impact and can serve as a basis for preventive measures. For each specific health outcome, one should consider not only the available factors that may modify the effect of noise but also the annoyance rating of noise sensitivity as the most important individual characteristic when predicting dissatisfaction with the noise [14, 17].

The health outcomes influenced by possible nonacoustic factors may include gender, age, education, subjective noise sensitivity, extroversion/introversion, general stress score, comorbidity, length of residence, duration of stay at dwelling in the day, window orientation of a bedroom or living room toward the street, personal evaluation of the source, attitudes toward the noise source, coping capacity with respect to noise, perception of malfeasance by the authorities responsible,

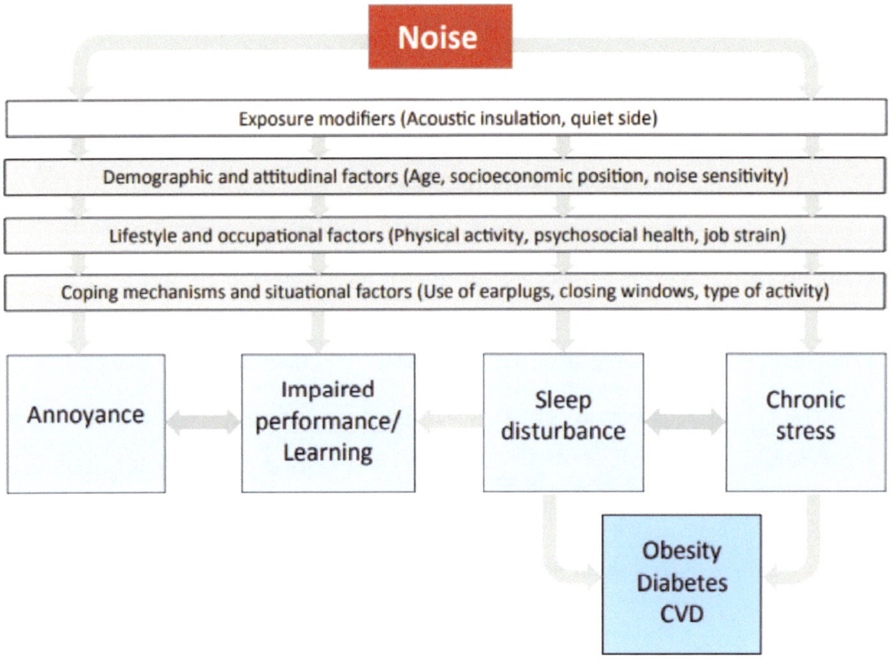

Figure 2.
The framework of health effects of noise according to environmental noise and health – current knowledge and research needs ISBN 978-91-620-6553-9.

body mass index, and smoking habits. In noise annoyance studies, nonacoustic factors may explain up to 33% of the variance [18].

According to WHO report, the key health outcomes associated with environmental noise exposure based on the seriousness and prevalence and the anticipated availability of evidence were in the following. The health outcomes were divided as either critical or important for developing recommendations on the health impacts of environmental noise. The selection of health outcomes was based on the available evidence for the association between the environmental noise and the specific outcome, as well as public concern about the health outcome resulting from noise exposure. The critical health outcomes associated with environmental noise included such as cardiovascular disease, annoyance, sleep disturbance, cognitive impairment, and hearing impairment and tinnitus. In addition, the important health outcomes were adverse birth outcomes, metabolic outcomes, quality of life, well-being, and mental health [19].

The following health outcomes were based on the evidence-based association between the environmental noise and the specific outcome, as well as public concern about the health outcome resulting from noise exposure. These health outcomes can be measured in various ways, and their prioritization was based on the impact of the disease and the disability weights (DWs) associated with the health outcome measure. A disability weight is a weight factor that reflects the severity of the disease on a scale from 0 (perfect health) to 1 (equivalent to death). Years Lost due to Disability (YLD) are calculated by multiplying the incident cases by duration and disability weight for the health condition [8]. In case of cardiovascular disease, DW of IHD is 0.405, DW of hypertension is 0.117, and the severity of the disease itself is high in IHD. However, the incidence rate varies depending on the survey area or country, the results of YLD may be different. The critical health outcomes, identification of the priority outcome measures, and justifications for their selections are listed in **Table 5** [19].

Critical health outcome	Critical health outcome measures	Justification for selection
Cardiovascular disease (L_{den})	Self-reported or measured prevalence, incidence, hospital admission, or mortality due to: • ischemic heart disease (IHD) (including angina pectoris and/or myocardial infarction) • hypertension • stroke	Except for self-reports, these are objective measures of the outcome, affect a large proportion of the population, have important health consequences, and can lead to more severe diseases and/or mortality. DW for IHD: 0.405 DW for hypertension: 0.117
Effects on sleep (L_{night})	• Percentage of the population highly sleep-disturbed (%HSD), self-reported, assessed with a standardized scale • Polysomnography measured outcomes (probability of additional awakenings) • Cardiac and blood pressure outcome measures during sleep • Motility measured sleep outcomes in adults • Sleep disturbance in children	This is the most meaningful, policy-relevant measure of this health outcome. Self-reported sleep disturbances are a very common problem in the general population: they affect quality of life directly and may also lead to subsequent health impediments. Effects on sleep may be in the causal pathway to cardiovascular disease. This measure is not a proxy for physiological sleep quality parameters but is an important outcome in its own right. DW for %HSD: 0.07
Annoyance (L_{den})	• Percentage of the population highly annoyed (%HA), assessed with standardized scale • Percentage annoyed, preferably assessed with standardized scale	This is the most objective measure of this health outcome. Large proportions of the population are affected by noise annoyance, even at relatively low exposure levels. Annoyance may be in the causal pathway to cardiovascular disease. DW for %HA: 0.02
Cognitive impairment (L_{den})	• Reading and oral comprehension, assessed with tests • Impairment assessed with standardized tests • Short- and long-term memory deficit • Attention deficit • Executive function deficit (working memory capacity)	This outcome measure is the most meaningful: it can affect vulnerable individuals (children) and has a significant impact later in life. DW for impaired reading and oral comprehension: 0.006
Hearing impairment and tinnitus (L_{Aeq} and $L_{AF,max}$)	• Permanent hearing impairment, measured by audiometry • Permanent tinnitus	This outcome measure can affect vulnerable individuals (children) and has a significant impact later in life. It is the most objective measure for which there is an ISO standard (ISO, 2013), specifying how to estimate noise-induced hearing loss. DW for mild severity level (threshold at 25 dB) for childhood onset: 0.0150

DW: A disability weight is a weight factor that reflects the severity of the disease on a scale from 0 (perfect health) to 1 (equivalent to death).

Table 5.
Critical health outcomes, outcome measures identified, and justifications for selection according to the WHO Environmental Noise Guidelines for the European Region.

6. The burden of environmental noise and adverse health outcome

The disability weight (DW) is used to rank the priority critical health outcome measures. DWs are ratings that vary between 0 and 1, in which 0 indicates no disability and 1 indicates the maximum amount of disability. The DWs have been proven useful in calculating the burden of disease (**Table 6**).

Priority health outcome measure (associated with DW)	Relevant risk increase considered for setting of guideline level
Incidence of IHD (DW: 0.405)	5% RR increase
Incidence of hypertension (DW: 0.117)	10% RR increase
%HA (DW: 0.02)	10% absolute risk
%HSD (DW: 0.07)	3% absolute risk
Permanent hearing impairment (DW: 0.0150)	No risk increase due to environmental noise
Reading and oral comprehension (DW: 0.006)	One-month delay in terms of reading age

Table 6.
Priority health outcomes and relevant risk increases for setting guideline levels according to the WHO Environmental Noise Guidelines for the European Region.

Important health outcome	Health outcome measures reviewed
Adverse birth outcomes (L_{den})	• Pre-term delivery • Low birth weight • Congenital anomalies
Quality of life, well-being, and mental health (L_{den})	• Self-reported health and quality of life • Medication intake for depression and anxiety • Self-reported depression, anxiety, and psychological distress • Interviewer-assessed depressive and anxiety disorders • Emotional and conduct disorders in children • Children's hyperactivity • Other mental health outcomes
Metabolic outcomes (L_{den})	Prevalence, incidence, hospital admission, or mortality due to: • type 2 diabetes • obesity

Table 7.
Important health outcomes and health outcome measures reviewed according to the WHO Environmental Noise Guidelines for the European Region.

For cardiovascular disease, the DW value (DW: 0.405) specifically applied to acute myocardial infarction in the publication outlining the data sources, 5% increase of relative risk in ischemic heart disease (IHD) and 10% in hypertension. The DWs for high sleep disturbance (DW: 0.07), high annoyance (DW: 0.02), and impaired reading and oral comprehension (DW: 0.006) were developed in the context of calculating the burden of disease from environmental noise. According to the WHO night noise guidelines, there were observed adverse health effects at levels starting from 40 dB L_{night}, and self-reported sleep disturbance (HSD) and annoyance should not exceed 3 and 10% to be health protective, receptively. The DW for hearing impairment was available from the technical paper on the burden of disease from environmental noise; a DW of 0.0150 for moderate severity level "has difficulty following a conversation in a noisy environment, but no other hearing problems." For cognitive impairment, the DW was derived from a very conservative value (DW: 0.006) for noise-related impairment of children's cognition, equivalent to a DW for contemporaneous cognitive deficit in the context of a range of cognitive impairments in children. This impact cannot be predicted accurately [9].

Also, WHO provides a list of the important health outcomes along with the reviewed measures. There was no prioritization of health outcome measures leading to justification of selection, since important health outcomes had less impact on the development of recommendations. In **Table 7**, the health outcome-related noise indicator was L_{den}, and the most common health outcomes were relevant to psychoacoustic problems such as quality of life, well-being, and mental health [20].

7. Mental health impacts of noise

Environmental noise is not believed to be a direct cause of mental illness, but it is assumed that it accelerates and intensifies the development of latent mental disorder. Studies on the adverse effects of environmental noise on mental health cover a variety of symptoms, including anxiety, emotional stress, nervous complaints, nausea, headaches, instability, argumentativeness, sexual impotency, changes in mood, increase in social conflicts, and general psychiatric disorders such as neurosis, psychosis, and hysteria [17, 21].

Then, even when exposed to lower noise levels than outdoor, why the problems of neighbor noise are taken so sensitively and seriously? This is because it is contrary to the expectation that home is a place of rest and a comfortable, quiet place.

New Yorkers, like citizens in the quietest towns of the country, expect less noise when they close the doors to their apartment and homes. They may willing to deal with the noisy street traffic, crowds, and subways, as they transverse the city but they are less tolerant of noisy intrusions into their homes (Why noise matter chap 2) [3].

It turned out that psychiatric disorders are associated with noise sensitivity, rather than with noise exposure level, and the association was found to disappear after adjustment for the baseline trait anxiety. These and other results show the importance of taking vulnerable groups into account because they may not be able to cope sufficiently with unwanted environmental noise. This is particularly true of children, the elderly, and people with preexisting illnesses, especially depression. Despite the weaknesses of the various studies, the possibility that community noise has adverse effects on mental health is suggested by studies on the use of medical drugs, such as tranquilizers and sleeping pills, on psychiatric symptoms and on mental hospital admission rates. About 1 of 10 people are particularly noise sensitive. These people will become 10% more annoyed by noise than general population [22].

After adjustment of noise-related variables, sociodemographic factors, medical illness, and duration of residence, subjects in the high noise-sensitive (NS) group were more than 2 times more likely to experience depression and insomnia and 1.9 times more likely to have anxiety, compared to those in the low NS group. The levels of noise recognition and psychological discomfort are affected by various factors, including individual components (e.g., age and effects of traits) and environmental factors, including contextual aspects and noise parameters (e.g., source, attitude toward noise, and amplitude modulation). Not all people exposed to environmental noise suffer from a disease or health problem, and the effects of noise differ among individuals [23].

Hearing noises above the perceived normal threshold, higher noise sensitivity, and continuous noises were associated with higher levels of displaced aggression (DA). It occurs when a person is provoked, is unwilling or unable to retaliate against the original provocateur, and subsequently aggresses against a seemingly innocent target. Low frequency and high intensity noises were also associated with higher DA scores. DA score was higher in women and in older people living in the neighborhood for a longer time, in people with better education, and in those reporting poorer health. Moreover, low frequency and continuous noises resulted in higher DA. The frequency of hearing noises above the normal threshold was positively correlated with DA as was noise annoyance [24].

8. Conclusion

The laws related to noise are increasingly being strengthened, and the actual noise level is decreasing compared to the past with the improvement of building technology and the reinforcement of regulatory standards. However, social problems related to noise have become more serious and are now an important part of the psychoacoustic problems related to the environment. This is also related to the expectation of a better quality of life as a result of the improvement of the economic level and the desire for home comfort. Also, even when the actual noise level is not high compared to other external environments and problems occur between neighbors. Therefore, it means that the problem of neighborhood noise should be discussed not only at the physical level but also at the psychoacoustic aspects, and the individual's sensitivity and cultural difference should be considered. In particular, it needs to be treated more seriously, considering that health problems (especially mental problems) related to neighborhood noise may occur at a lower level than actual measured noise levels. Also, neighborhood noises tend to provoke the existing mental health problems related to noise more easily. Identifying the physiological and psychological effect of environmental noise precisely and making it recognized broadly comprise the essential part of solving the environmental noise problem to create better living environment. The neighborhood noise issues were in part due to the lack of communication between and among neighbors. Lack of contact within the community affected people's perception of the loudness of daily life sounds [6]. Furthermore, intense emotional conflicts between and among some neighbors were also found to be one of the major sources of daily life noise. Even though the neighborhood noise annoyance will always be an issue, we want to be good neighbors and also hope to have good relationship with neighbors. Consideration of neighborhood, social responsibility, and changes in social behavior are important factors in addressing neighborhood noise. In addition, it is necessary to understand a scientific basis for the health impact of noise and to provide national support for noise reduction.

Conflict of interest

The author declares no conflict of interest.

Thanks

I would like to thank the members of the Department of Occupational and Environmental Medicine, Ulsan University Hospital. A Ram Kim, Daeyun Kim, and Sunghee Lee have contributed directly or indirectly to this chapter. We have shared information and ideas. Moreover, they have made suggestions and comments.

Author details

Jiho Lee
Department of Occupational and Environmental Medicine, Ulsan University Hospital, University of Ulsan College of Medicine, Ulsan, Korea

*Address all correspondence to: oemdoc@naver.com

References

[1] Locher B, Piquerez A, Habermacher M, Ragettli M, Roosli M, Brink M, et al. Differences between outdoor and indoor sound levels for open, tilted, and closed windows. International Journal of Environmental Research and Public Health. 2018;**15**(1):149

[2] WHO. World Health Organization International Program on Chemical Safety. Géneve: Executive Board: Provisional Agenda Item 12; 1994

[3] Stewart J, McManus F, Rodgers N, Weedon V, Bronzaft A. Why Noise Matters: A Worldwide Perspective on the Problems, Policies and Solutions. London, New York: Routledge; 2016

[4] Bronzaft A. Neighborhood Noise and its Consequences. New York: Survey Research Unit, School of Public Affairs, Baruch College; 2007

[5] Conde-Sala JL, Portellano-Ortiz C, Calvo-Perxas L, Garre-Olmo J. Quality of life in people aged 65+ in Europe: Associated factors and models of social welfare-analysis of data from the SHARE project (wave 5). Quality of Life Research. 2017;**26**(4):1059-1070

[6] Yoshioka-Maeda K. A preliminary review of literatures focusing on the neighborhood noise issue in Japan. Asian Pacific Journal of Disease Management. 2014;**8**(1-2):1-8

[7] Babisch W. Transportation noise and cardiovascular risk: Updated review and synthesis of epidemiological studies indicate that the evidence has increased. Noise & Health. 2006;**8**(30):1-29

[8] Organization WH. Burden of Disease from Environmental Noise: Quantification of Healthy Life Years Lost in Europe: World Health Organization. Copenhagen: Regional Office for Europe; 2011

[9] Clark C, Paunovic K. WHO environmental noise guidelines for the European region: A systematic review on environmental noise and cognition. International Journal of Environmental Research and Public Health. 2018;**15**(2):285

[10] Gjestland T. A systematic review of the basis for WHO's new recommendation for limiting aircraft noise annoyance. International Journal of Environmental Research and Public Health. 2018;**15**(12):2717

[11] WHO. Night Noise Guide Lines for Europe. Copenhagen: WHO Regional Director for Europe; 2018. pp. 17-19

[12] Basner M, McGuire S. WHO environmental noise guidelines for the European region: A systematic review on environmental noise and effects on sleep. International Journal of Environmental Research and Public Health. 2018;**15**(3):519

[13] Basner M, Babisch W, Davis A, Brink M, Clark C, Janssen S, et al. Auditory and non-auditory effects of noise on health. Lancet. 2014;**383**(9925): 1325-1332

[14] Guski R. Personal and social variables as co-determinants of noise annoyance. Noise & Health. 1999;**1**(3):45-56

[15] Kim A, Sung JH, Bang JH, Cho SW, Lee J, Sim CS. Effects of self-reported sensitivity and road-traffic noise levels on the immune system. PLoS One. 2017;**12**(10):e0187084

[16] Eriksson C, Nilsson ME, Pershagen G. Environmental Noise and Health: Current Knowledge and Research Needs. Stockholm: The Swedish Environmental Protection Agency; 2013

[17] Omlin S, Bauer GF, Brink M. Effects of noise from non-traffic-related

ambient sources on sleep: Review of the literature of 1990-2010. Noise & Health. 2011;**13**(53):299-309

[18] Guski R, Schreckenberg D, Schuemer R. WHO environmental noise guidelines for the European region: A systematic review on environmental noise and annoyance. International Journal of Environmental Research and Public Health. 2017;**14**(12):1539

[19] Clark C, Paunovic K. WHO environmental noise guidelines for the European region: A systematic review on environmental noise and quality of life, wellbeing and mental health. International Journal of Environmental Research and Public Health. 2018;**15**(11):2400

[20] Clark C, Crumpler C, Notley AH. Evidence for environmental noise effects on health for the United Kingdom policy context: A systematic review of the effects of environmental noise on mental health, wellbeing, quality of life, cancer, dementia, birth, reproductive outcomes, and cognition. International Journal of Environmental Research and Public Health. 2020;**17**(2):393

[21] Berglund B, Lindvall T, Schwela DH, WHO. Guidelines for Community Noise. Occupational and Environmental Health Team; 1999

[22] Maris E, Stallen PJ, Vermunt R, Steensma H. Noise within the social context: Annoyance reduction through fair procedures. The Journal of the Acoustical Society of America. 2007;**121**(4):2000-2010

[23] Park J, Chung S, Lee J, Sung JH, Cho SW, Sim CS. Noise sensitivity, rather than noise level, predicts the non-auditory effects of noise in community samples: A population-based survey. BMC Public Health. 2017;**17**(1):315

[24] Dzhambov A, Dimitrova D. Neighborhood noise pollution as a determinant of displaced aggression: A pilot study. Noise & Health. 2014;**16**(69):95-101

Chapter 3

Attenuation of Environmental Noise through Digital Filtering

Juan Ríos and Celedonio Aguilar

Abstract

This chapter proposes the implementation of an environmental noise cancelation system using the least mean squares algorithm with noise amplitude modulation (NAMLMS). The system was implemented in a dsPIC3020F10 digital processor. The results obtained demonstrate that low- and high-frequency signals are attenuated allowing the passage of an audible range between 5 kHz and 18.9 kHz, the above using real-time processing.

Keywords: noise, amplitude, attenuation, digital, filtering

1. Introduction

It is important to consider that, after vision, the ear is the most important sensory organ of the human body. The ear is an alarm sensor; that is, it is always active to perceive sounds. The sound consists in the change of air pressure that produces waves with a certain periodic frequency. However, within these there are also sound signals that are inarticulate to the human ear and also unpleasant, even damaging human health. Any sound pressure in the air is measured in decibels. A decibel is a logarithmic value (not linear, but exponential) that represents the relationship between a measured value and a reference value. In any environment there are audible signals that we need to perceive (talks, alarms, etc.) and other unwanted signals that cause the so-called noise pollution. The said contamination is defined as the presence of noise caused by an acoustic emitter that implies discomfort or harm to the health of people.

Some of the noises to which the human auditory system is exposed and that are part of the noise pollution are the following:

1. Noises due to the use of musical instruments
2. Use of household appliances such as washing machines, vacuum cleaners, etc.
3. Noises of air conditioning equipment
4. Discomfort caused by infrastructure
5. Dogs barking in the absence of their master
6. Alarms and sirens caused by fixed sources and mobile sources

Noise pollution significantly interferes with interpersonal communication and increases workplace accidents and traffic accidents. **Table 1** shows the noise levels and their effects on the human ear.

A filter may consist of hardware or software applied to a set of noisy or contaminated data in order to extract information of interest. Noise can be generated by any source. In this chapter, a digital solution is proposed that allows the attenuation of noise attached to audible signals that are perceived by the human ear. An adaptive filter is a device that is useful for processing an input signal by blocking and/or allowing some parts of it (unwanted noise or vibration).

On the other hand, the adaptive filter [1] has a feedback loop that has the ability to minimize the error produced by the comparison between the output signal (after processing) and a reference signal (expected signal). The above allows to guarantee that the human ear is not contaminated with the unwanted signals and only to rescue the information that needs to be received. **Figure 1** shows the structure of an adaptive filter.

The signal d(n) is the expected audio signal to be perceived by the human ear, x (n) is the input signal to the filter (what is perceived from the environment), y(n) is the response signal that grants the filter, and e(n) is the error required to adapt the filter parameters by comparing the desired signal with that obtained. This comparison is the difference between the desired (reference) signal with respect to the signal obtained at the filter output.

dBa	Characteristics
130	It is the level perceived at approximately 10 meters away from an airplane. The noise is unbearable and painful.
120	It is dangerous, ear protection is required. This noise is produced by the reactor of an airplane about 50 meters away.
110	They are dangerous and very annoying. They are habitual in a disco, at a rock concert and 100 meters from a plane landing.
90	It represents a very noisy environment and is dangerous if exposure occurs for a long time. It is the characteristic noise level of a heavy vehicle traveling at 60 km/h and perceived at about 10 meters
80	It corresponds to quite noisy environments, such as a street with heavy traffic or some appliances such as vacuum cleaners or washing machines.
70	It represents a noisy environment, habitual in comercial areas and many bars, inside a train or a car.
60	It is equivalent to a little noisy environment and is the usual level of voice sound in a normal conversation.
50	It represents a quiet environment, although they still interfere with sleep. It is the usual level in a study room.
40	It represents a calm environment and admissible to maintain sleep.
30	It represents a quiet environment.

Table 1.
Noise levels and their effects.

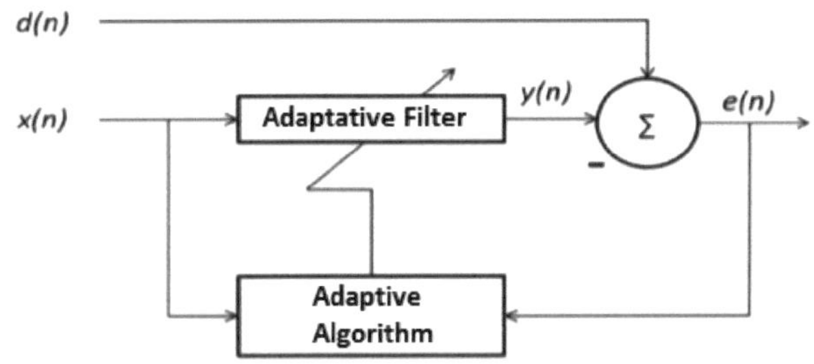

Figure 1.
Generic structure of an adaptive filter.

The operation of the algorithm that will represent the filter [2] is as follows:

1. The input signal will be picked up (in the scheme it is labeled as x(n)). This signal is the one that is in the environment and that contains what you want to be perceived and the adhered noise.

2. There is a reference signal labeled with d(n). This signal is what is expected to be perceived by the human ear, for example, the conversation you have with the person with whom you are talking.

3. The signal x(n) is processed in order to attenuate unwanted signals (noises that may be high or low frequencies) in order to obtain y(n).

4. The output signal y(n) and the desired d(y) are compared in order to obtain an error that should be the minimum possible.

5. Based on the error generated, the process will be carried out again to ensure that the signal obtained after the filter is as close as possible to the reference signal.

In general, the filter [3] will be used to inhibit signals that are not desired (technically this process is known as attenuating the amplitude of the noise signal) to "clean" the signal that must be perceived by the human ear. The proposed technique adapts to the conditions of the environment in which it is being implemented.

Section 2 shows the development of the proposal; the equipment required for its implementation and its constituent parts are defined.

2. Design of the proposed algorithm and scheme

The design of the proposal will be partitioned in stages. Each stage involves material and equipment required for the tests, measurements, and results obtained. **Figure 2** shows the process diagram that will follow the treatment of the signal captured from the environment.

The general processing diagram shown in **Figure 2** consists of five stages which will be explained below:

Stage A. It consists of the acquisition of the audible signal that is recovered from the environment. The use of an omnidirectional microphone is feasible here, the foregoing because the audio signal can be obtained from any point but not from a specific one. A microphone is a transducer device that converts sound waves into electrical voltage changes. In simple words, the microphone will record all the signals captured from any side from where they are received since the entire microphone structure is sensitive to sound. Next, **Figure 3** shows the pickup pattern of an omnidirectional microphone.

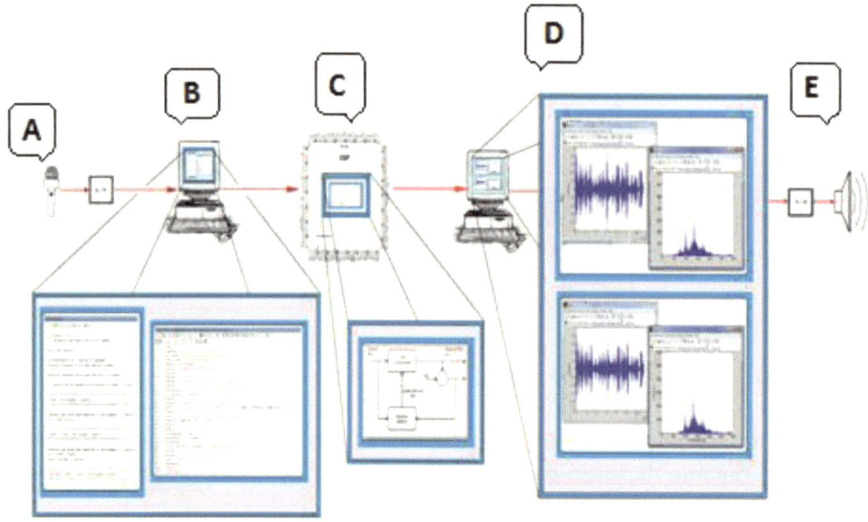

Figure 2.
General processing diagram.

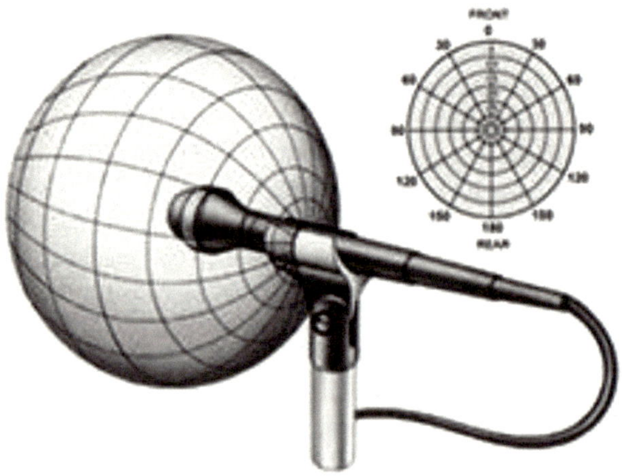

Figure 3.
Omnidirectional pickup pattern in a microphone.

It should be mentioned that an omnidirectional pickup pattern captures the sound obtained from any direction.

It is important to capture all the sounds of the environment. For example, if a person is going to cross the street, it is important that he consider the sounds of fenced cars and be attentive to all perceived sounds in order to avoid an accident. The proposal to be designed will minimize the sound intensity of these noises but not completely mitigate them.

The proposal will capture all the random noises in the environment and process them, and when it detects that there is one that exceeds the threshold of intensity allowed, it will multiply it by a reduction factor and thus decrease its volume. The process described above is purely adaptive, which means that "n" number of repetitions will be performed until it approaches the expected result.

Stage B. The signal captured in **Stage A** is of analog format since it was captured in its natural form. However, for this stage, it is necessary to convert the captured signal into digital format since the computer and the processor require scanning to perform its function. The original signal has to be subjected to three basic operations to digitize it: sampling, quantization, and coding.

Sound, by nature, is an analog signal. It is produced by vibrations in the air that force the union of nearby molecules in the air by slightly raising its pressure. Such pressure changes reach the ear by vibrating the receptors and decoding to produce the sound. Some of the characteristics of the vibrations (in waveforms) are the following:

a. Amplitude: reflects the change in pressure from the highest peak to the minimum. A waveform with large amplitude has a volume of equal magnitude; otherwise, the volume is quieter.

b. Cycle: describes a single repeated sequence of pressure changes.

c. Frequency: describes the number of cycles produced in a second. If the frequency is high, the tone of the sound will be higher.

Figure 4 shows the representation of two analog signals of different frequencies each.

If the frequency of a sound signals increases, the sound will be perceived as more acute because the wavelength decreases. Otherwise, when the frequency is low, the repetitions decrease, the wavelength increases, and, therefore, the sound tends to be a high-pitched tone.

The first step to digitize the original signal consists of the sampling operation. The sampling of a sound signal consists of taking small representative pieces of the signal so that they are then encoded in binary digits to digitize them.

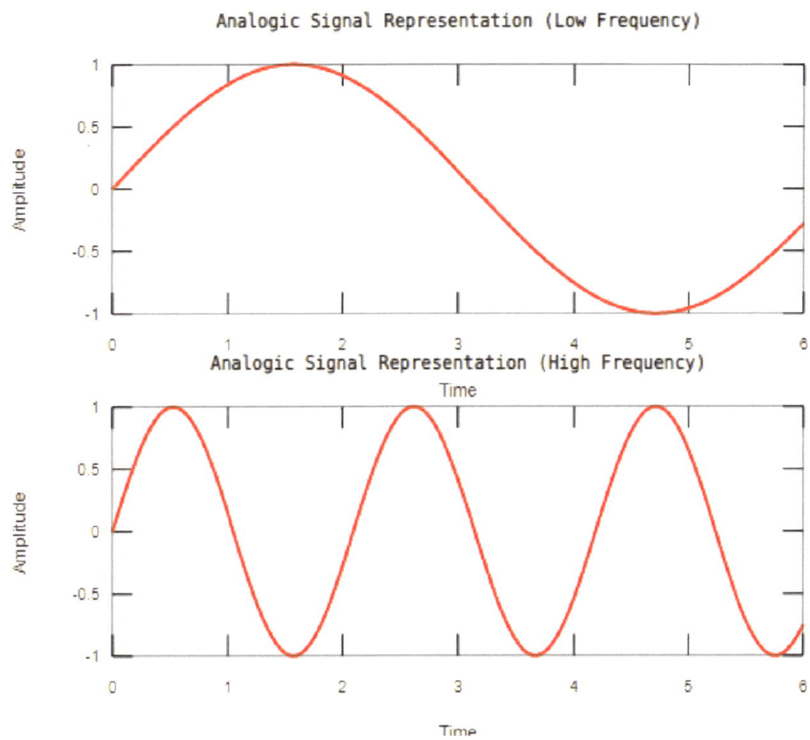

Figure 4.
Analogic signals with different frequencies.

The condition that the signal has to remain representative of the original must be considered. To cover the previous condition, the following equation known as Nyquist's theorem must be followed:

$$fs \geq 2fo \tag{1}$$

It is worth mentioning that if the sampling frequency is high, that is, if more samples are taken from the original signal in a certain time, the collection of significant parts for later digitization will allow a better fidelity of the original signal but now in a digital format and ready for computational processing.

Figure 5 represents the two signals shown in **Figure 4** but now applying the sampling theorem; that is, small samples (pieces) representative of the signal are taken to be the "objects" to which binary codes will be assigned and thus digitized.

The quantization is the second step. It consists in assigning amplitude values to each sample obtained in the previous step, the foregoing in order to identify each sample that will be encoded.

Finally, and based on the amplitude assigned to each sample, a binary combination is correlated to each of them to be identified. Samples that have the same amplitude will have the same binary code.

Stage C. In this step the adaptive filter algorithm will be implemented in the dsPIC3020F10 in order to perform the tests in real time. The processor will receive the signal in digital format (under previous scanning in **Stage B**) in order to process it to block the noise and allow the desired signal to pass through.

A DsPIC is a type of microprocessor known as a digital signal processor. It is responsible for real-time processing, a feature that is essential when non-tolerance of delays is required. Basically, a DsPIC [4] acquires a digital signal and processes it to improve it (in the case of audio, a clearer and sharper sound).

The algorithm to be implemented is the noise amplitude modulation least mean square (NAMLMS) which consists in a modification of the LMS algorithm, the

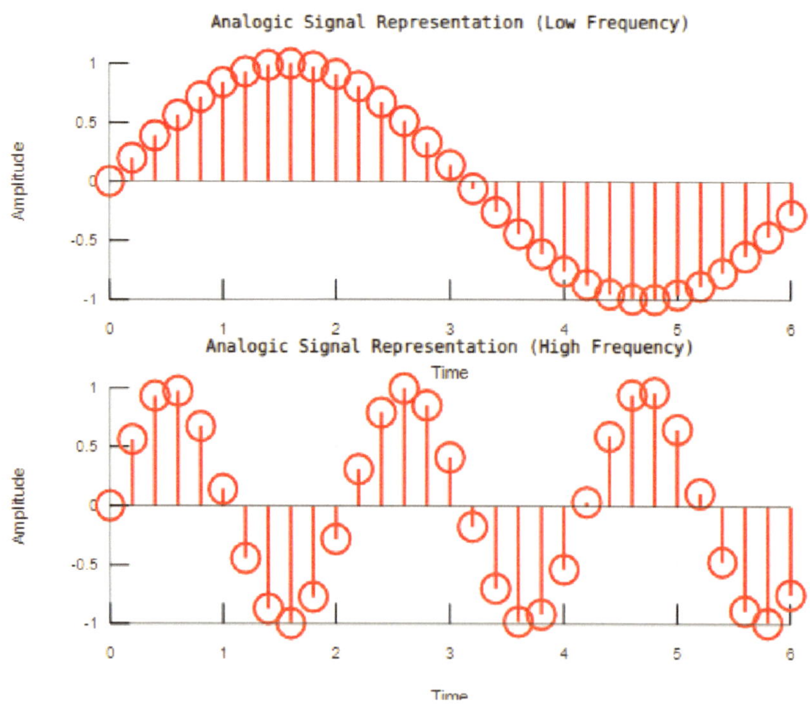

Figure 5.
Sampled analogic signals.

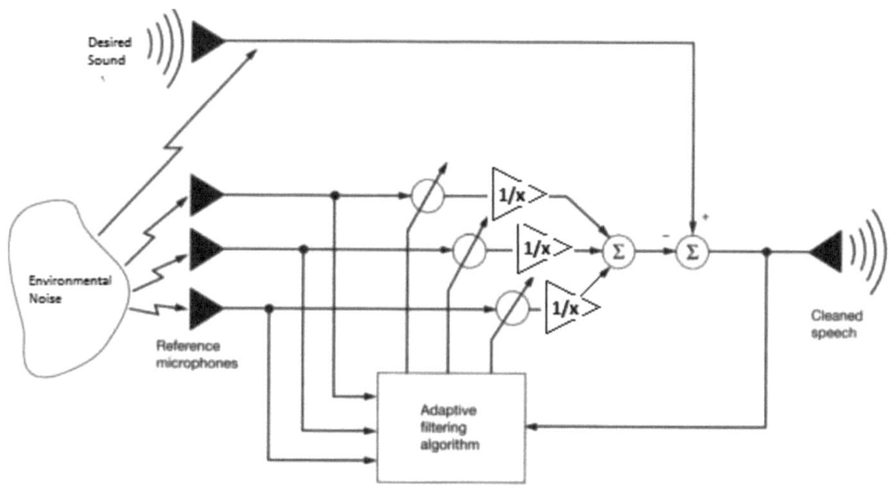

Figure 6.
Scheme of the proposal to cancel environmental noise.

foregoing because the noise will have a modulation in its amplitude in order to eradicate it as well as possible.

There is a wide variety of algorithms that can be classified into algorithms of low computational complexity with low convergence speed and high convergence speed algorithms with high computational cost. Some of the most used algorithms due to their low computational complexity are the averaged least squares algorithm (LMS, least mean square) and its normalization (NLMS). These algorithms have been successfully implemented in different systems. However, the convergence speed of these algorithms is slow. This means that the processing turns out to be so slow that it reduces the tests in real time, which would be not feasible for noise reduction applications especially in people suffering from hearing loss.

The proposed scheme of **Figure 6** consists of the following stages for the process:

1. Detection of the desired signal mixed with the ambient noise.

2. Attenuation of the environmental noise signal at a rate of 1/x, in order to reduce its amplitude and thus mitigate it. The above will happen only if the signal amplitude exceeds the allowed threshold. The frequency of it is not a factor to consider because we do not intend to distort the noise, only adaptively manipulate its sound intensity.

3. The desired signal is subtracted from the amplitude-modulated noise signal.

4. The signal obtained after the process is compared with respect to the desired one in order to obtain an error.

5. Based on the error obtained, the system is adapted to decide whether to reperform the process or has already converged to the permissible and minimum tolerated error.

Next, the process to follow for the mathematical analysis of the system is shown:

1. The desired signal with the adhered noise is picked up by the system. It is worth mentioning that the desired sound signal is contaminated by the noise prevailing in the environment:

$$s(n) = d(n) + \frac{1}{x} v(n) \tag{2}$$

where
s (n) = signal captured;
d (n) = desired signal;
v (n) = adhered environmental noise; and
x = modulation coefficient for attenuation.

It is worth mentioning that the modulation coefficient will also be submitted to the adaptation algorithm in order to serve to attenuate the noise signal generated in the environment.

2. The correlation between both signals must be equal to zero. The above explains the fact that both are linearly independent:

$$r_{sv} = \frac{cov(s, v)}{cov(s) * cov(v)} = 0 \tag{3}$$

3. Subsequently, the signal obtained is processed by the adaptive filter to produce the output:

$$y(n) = \sum_{k=0}^{M-1} w_k(n) * v(n - k) \tag{4}$$

where
w (n) = the values of the adjustable coefficients of the adaptive filter; and.
k = iteration for each adaptation.

4. The filter output y (n) is subtracted from the main signal s (n). The above defines the error signal:

$$e(n) = s(n) - \frac{1}{x} v(n) \tag{5}$$

5. The error signal is the one used to adjust the coefficient values of the adaptive filter and control loop around filtering operations and subtraction are related. Minimizing the mean square value of the error signal means maximizing the signal to noise ratio of the system output.

6. The adaptive filtering operation is perfect when:

$$y(n) = d(n) \text{ and with} e(n) = 0 \tag{6}$$

In this case, the system output is noise-free, and the noise cancelation is perfect. Correspondingly, the signal to noise ratio of the output is infinitely large.

Stage D. It consists of validating the filter response and also verifying that it inhibits unwanted frequencies (adhered noise). It is important to consider that the frequencies that will correspond to the desired signal must be the sounds that are desired and everything else must be blocked.

Stage E. It is the inverse process of **Stage B**. The sound has an analog nature, so it must be converted to that format so that the receiver can decode it.

3. Results of the implementation

In the environment, the desired sound signals are mixed with the unwanted signals (noise) [5]. **Figure** 7 shows the acquisition of the desired signal and the mixed noisy signal.

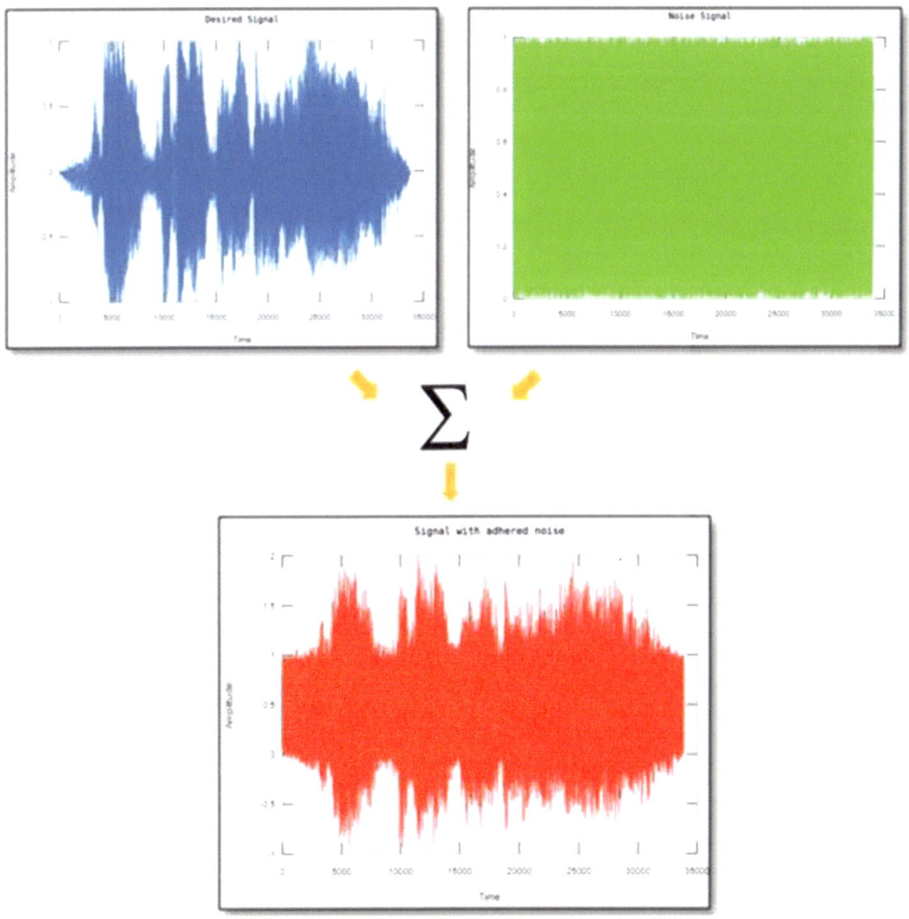

Figure 7.
Combination between the original (desired) signal and the noise signal.

The blue graph shows the desired audio signal and used as a reference to compare what you want the system to throw at the output. The green signal is a type of noise captured in the environment, it could be said to be a random signal since it does not have a defined pattern. Finally, the red signal is the mixture of the previous ones.

The objective is to submit the mixed signal (red) to the proposed adaptive system in order to attenuate the noise signal and allow the sound signal to pass through.

The designed filter adapts to the conditions of the environment in which it is implemented. Next, the following figures will show, in parts, the analysis obtained for the proposed algorithm and its comparison with others established in the existing literature related to the subject.

Figure 8 shows the frequency response obtained for the passage of frequencies above the 10 kHz frequency which, on average, is the frequency at which a desirable sound oscillates. It is worth mentioning that, as a result of the adaptation, there is a slope that could establish a tolerance margin at lower frequencies.

As can be seen in **Figure 8**, the response of the NAMLMS algorithm shows a slight hesitation in stability but achieves greater convergence with respect to the responses of the other algorithms. The proposed algorithm was based on the LMS algorithm that results in obvious instability and convergence, concluded, but showing latency.

Figure 9 shows the response for the part of the filter that allows low frequencies to pass through. The comparison with the responses of other filtering algorithms is

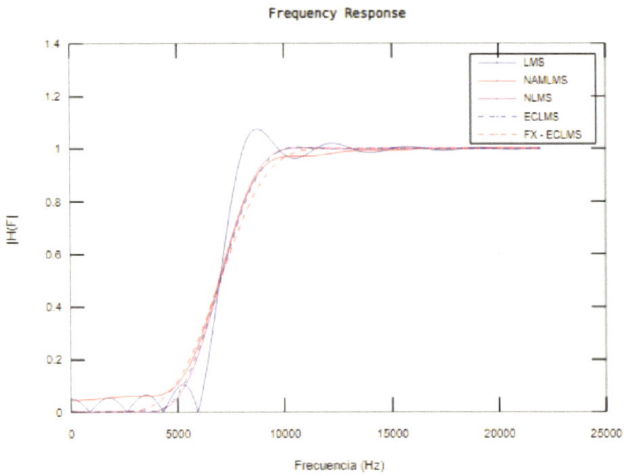

Figure 8.
Frequency response of the high-pass filter.

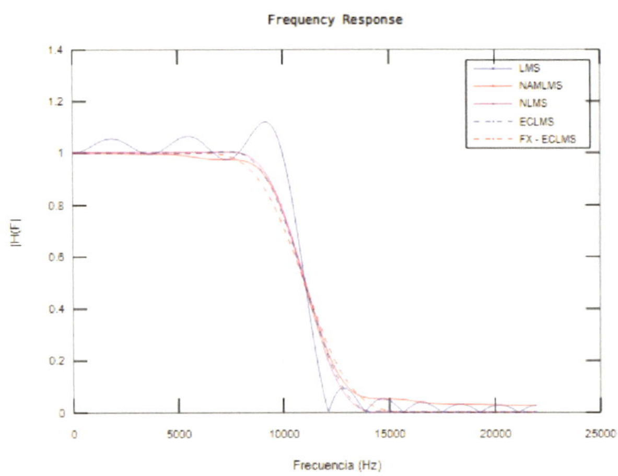

Figure 9.
Frequency response of the low-pass filter.

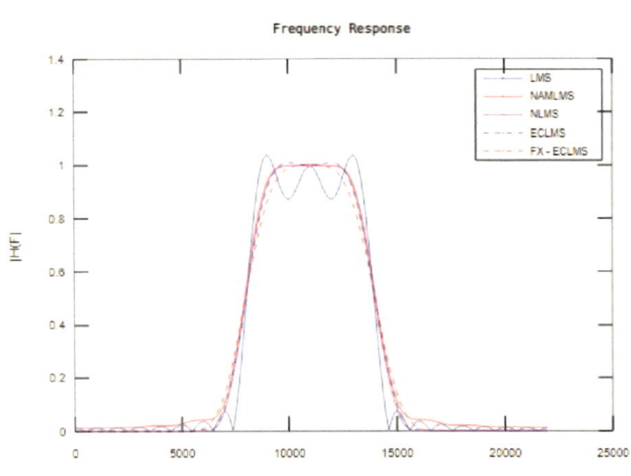

Figure 10.
Frequency response of the band pass filter.

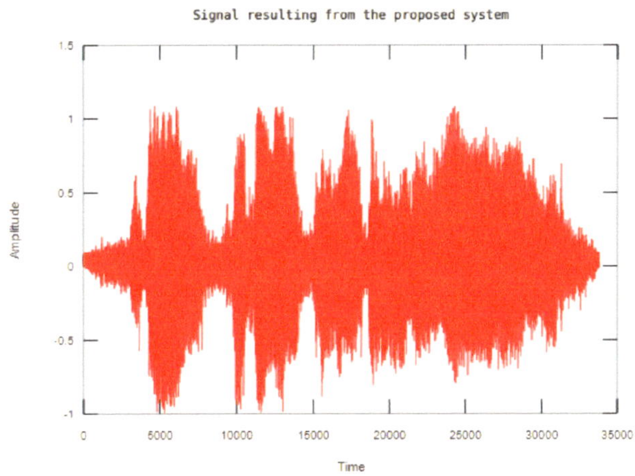

Figure 11.
Signal resulting.

visualized, and the rapid convergence and adaptation of the response offered by the proposed system stands out.

Figure 10 shows the response of the system in general. It can be seen that the union of the two previous frequency responses generates the response of a frequency range through the filter. The objective is to suppress very low frequencies and very high frequencies that can be considered as noise and distortions that affect the human ear causing hearing loss.

Figure 11 shows the signal obtained as a response from the system. When compared with the reference signal, a correlation value of 0.912 is calculated which, according to the theory, indicates that there is a strong correlation between both signals and that, although the noise is slightly perceived, the desired signal is clearly perceived.

4. Conclusion

This chapter proposes an ambient noise cancelation system that allows to attenuate the noise that is mixed with the desired audible signals. Based on the results obtained, it is verified that the convergence of the algorithm is rapid relative to other existing ones, so it can be useful for use in new-generation cochlear implants or as treatments against symptoms of hearing loss.

Subjectively, the algorithm has a slight perception of adhered noise but does not affect its acoustic apparatus or the decoding of the messages of the desired sound signal.

Objectively, the correlation between the signal obtained after the respective system was calculated to the reference signal (desired), and an almost perfect result was achieved. Obviously a 100% correlation is not possible because the noise adhered to the desired sound and it is impossible not to modify some samples of said signal.

Acknowledgements

We would like to thank the Universidad del Valle de México for providing us with the necessary means to carry out the research and provide part of the resources required.

On the other hand, we also thank the Instituto Politécnico Nacional that has provided us with its facilities for experimentation and subject to real tests of the system I proposed. Without this valuable help, the development of this work would have been practically impossible.

Author details

Juan Ríos[1]* and Celedonio Aguilar[2]

1 Universidad del Valle de México, Mexico City, Mexico

2 Instituto Politécnico Nacional ESIME, Mexico City, Mexico

*Address all correspondence to: juan.riosc@uvmnet.edu

References

[1] Bhupendra KA, Miles AC, Chong H. A new adaptive algorithm and its implementation in MOS LSI. IEEE Journal of Solid-State Circuits. 1979; **14**(4):747-753

[2] Bhupendra KA, Miles AC, Chong HC. A sampled analog MOS LSI adaptive filter. IEEE Journal of Solid-State Circuits. 1979;**14**(1):148-154

[3] Bogucka H, Wesolowski K. Frequency-domain Echo cancellation in digital multicarrier modulation systems. IEEE Transactions on Communications. 2000;**48**(2):333-342

[4] Ávalos JG, González JM, Velázquez J, Sánchez JC. Implementación del algoritmo LMS con error codificado en el DSP TMS320C6713. In: Congreso Nacional de Ingeniería Electromecánica y de Sistemas, Ciudad de México, México, Noviembre 26–30, 2007; 2007

[5] Garcia M, Diego P, Quintana R: DSP implementation of the FxLMS algorithm for active noise control: Texas instruments TSM320C6713DSK. In: Automatic Control (CCAC), IEEE 2nd Colombian Conference; 2015. pp. 1-6

Chapter 4

Innovative Approaches to Noise Reduction

Mia Suhanek and Sanja Grubesa

Abstract

Nowadays, each individual is exposed to noise on a daily basis, and noise is often referred as in literature as a plague of modern society. Noise pollution is often overlooked when compared to other environmental pollutions (e.g. air, water, soil pollution). However, same as the all aforementioned pollutions, noise exposure has an accumulating character, meaning that the harmful effect of noise is detected only after a long period of time. Long exposure to noise pollution can be displayed as a bad mood, fatigue, insomnia, headache and loss of concentration, which causes reduced work ability and ultimately permanent hearing impairment. The goal of this chapter is to present two different approaches (traditional and contemporary) in noise reductions. The aim of both approaches is to link objective and subjective acoustic parameters, in order to plan future urban infrastructures while keeping in mind the existing acoustic environments, and to create and implement new solutions that will design, preserve and improve acoustic environments. Thus, we can conclude this chapter will be oriented towards human health and overall quality of life in terms of noise reduction.

Keywords: noise, noise pollution, noise reduction, acoustic parameters, noise barriers, soundscape

1. Introduction

Noise pollution is defined as any disturbing or unwanted noise that affects or deteriorates human or wildlife. Although noise constantly surrounds us, noise pollution generally receives less attention than, for example, water quality and air quality concerns, because it cannot be seen, tasted or smelled. Nonetheless, it is an indisputable fact that noise has a negative impact on everyday life especially if we observe urban areas. This chapter presents two ways of dealing with noise pollution in terms of reducing its levels. One is a more traditional approach and the other a more modern and very popular today with many directions and ways of implementations. The first major section will describe noise barriers with all their advantages and disadvantages, while in the second major section, soundscape approach will be discussed. The common goal of both approaches and of this chapter is to address the problem of noise pollution while bearing in mind the overall quality of life especially in urban areas.

2. A traditional approach: noise barriers

In order to reduce noise pollution, different protection measures can be applied. In terms of traffic noise pollution, reducing the impact of traffic noise on both people and the environments can be achieved by planning and integrating the traffic routes outside the residential areas. In case of existing traffic routes within the residential areas, a good solution for reducing the noise levels is noise barriers [1–3]. Here we note that the noise barrier efficiency depends principally on their design. In the field of noise barriers, it is already established that the most favourable noise barriers are those which have a diffuse element on the top. In addition, the diffuse element can be circular, is Y or T shaped and is usually added on the top of the plain barrier. In particular, the Y and T shapes have proven to be a very good choice for the shape of the diffuse elements [4–7]. Ishizuka and Fujiware [4] gave an extensive overview of the acoustic efficiency for several typical diffuse element forms placed at the top of the noise barrier. **Figure 1** shows a plane (reference) noise barrier and several other noise barrier types obtained by adding capes at the top of noise barriers bearing in mind that the caps are made of different materials [4]. The optimization of T-shape noise barriers was more thoroughly studied by Baulac et al. [5] and Monazam and Lam [6], while Grainer et al. [7] explored the Y-shape noise barrier optimization.

In Toledo et al. [8] a procedure was proposed for improving the acoustic efficiency of noise barriers using top-edge devices. Furthermore, in Toledo et al. [9] a procedure was developed for the optimization of well-based designs on the top of road barriers with both thick and very thin bodies by coupling a genetic algorithm with a 2D Dual BEM code. In addition, when placing a noise barrier in residential areas, studies have shown that it is also essential to keep in mind the "visual pleasantness" of the noise barrier which is the parameter introduced in Maffei et al. [10, 11].

Grubeša et al. [12, 13] have addressed the problem of economic feasibility of building noise barriers of various shapes and materials. Research and calculations done in this paper suggest a new specific noise barrier cost parameter (Ke) that must be taken into account during the optimization process of noise barrier shapes and materials while using computational calculations and optimization methods.

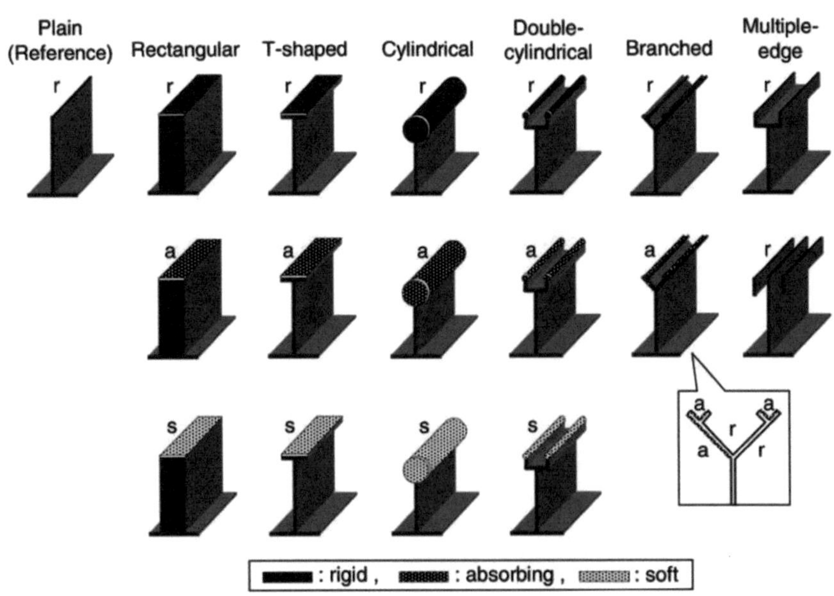

Figure 1.
Different types of noise barriers [4].

2.1 Noise level reduction with noise barriers

There are three basic parameters which describe noise barriers: insertion loss (IL), transmission losses (TL) and barrier absorption coefficient. Noise barriers can be defined as a certain sound "obstacle" between the sound source and the observer, i.e. the sound propagates around and over noise barriers. However, in real-case scenarios, the sound propagates also through the noise barrier, which is usually neglected, i.e. the sound proportion passing through the barrier is substantially smaller than the sound proportion which will cross over and around the noise barrier. The noise level reduction achieved by the installation of noise barriers is often called an additional noise level reduction, since the noise level will be primarily reduced due to the distance from the source and the air absorption and furthermore because the noise barrier itself. When quantifying noise reduction, a parameter, entered losses, is also often used and is defined as the noise level reduction arisen from the installation of noise barrier (insertion loss). It represents the difference between the sound pressures p_p and p_n, which are measured at the observer location before and after the noise barrier is placed, with the same ground configuration and position of the source and receiver, calculated according to the expression in Eq. (1). This parameter is usually used for comparison of different noise barrier performances:

$$IL = -20\, log\, log\left(\frac{p_p}{p_n}\right) \tag{1}$$

The noise reduction parameter which arises from the installation of the noise barrier depends on the shape and material of the noise barrier, the frequency and type of sound source, the position of the noise barrier with respect to the source and the observer and the absorption properties of the soil on both sides of the noise barrier. The noise barrier effectiveness directly depends on the frequency of the sound propagating over it, and therefore the parameter insertion loss (IL) is also frequency dependent.

The noise barrier sound-absorbing properties are qualified according to EN 1793-1 [14], while the airborne sound insulation index which corresponds to the transmitted noise barrier losses is defined in EN 1793-2 [15].

2.2 Noise barrier types

There are three basic types of noise barriers. These are ground-mounted noise barriers, structure-mounted noise barriers and a combination of the first two types. Ground-mounted noise barriers are constructed of natural earth materials such as earth, stone, rocks or gravel. This type of noise barriers is typically constructed from excess materials in a noise-protected location, and source and availability of such natural materials are factors that can significantly affect the cost of such noise protection. Ground-mounted noise barriers take up more space than structure-mounted noise barriers. The reason for this is the slope of the embankment, which must gradually increase in order to maintain the stability of the whole structure. The increase is defined by the ratio m:n, where m is growth in the horizontal direction and n is growth in the vertical direction. For most embankments, the ratio is 2:1 or 1.5:1, while for stone embankments, the increase is usually 1:1.

Structure-mounted noise barriers or commonly called just noise barriers can be:

- Panel, shown in **Figure 2**.
- Brick and masonry, shown in **Figure 3** [16].

Figure 2.
Panel noise barriers.

Figure 3.
Brick and masonry noise barriers.

- Freestanding which can be:
    - Cast concrete at the installation site, shown in **Figure 4** [16].
    - Concrete blocks manufactured under controlled conditions then delivered and positioned at the installation site, shown in **Figure 5**.
    - Green vertical gardens, shown in **Figure 6**.

2.2.1 Panel noise barriers

Panel noise barriers usually consist of a board or panel, which can be wooden, metal or concrete, and it can be constructed out of one piece, or it can be assembled at the place of noise barrier installation from several components. The panels are

Figure 4.
Cast-in place noise barriers.

Figure 5.
Precast concrete noise barriers.

Figure 6.
Green vertical noise barriers.

mounted between the base posts. The basic elements of this noise barrier type are the post and the elements which attach it to the foundation, the panels and the elements which attach the panels to the posts.

There are several ways to set up or build a foundation for posts:

- Reinforced concrete foundation with a post anchored to the top of the foundation using anchor bolts.
- Reinforced concrete foundation where the post is partially embedded in the concrete mass during concreting.
- Continuous foundation wall.
- Unreinforced concrete foundation with post submerged to full depth of foundation.
- Wooden posts dug into drilled cylindrical holes with stone fill.

2.2.2 Brick and masonry noise barriers

Brick and masonry noise barrier units are constructed out of either finished brick or masonry made of precast concrete blocks. Both types of noise barriers can be built at the installation site (by hand or machine), or they can be prefabricated in the form of blocks which will be assembled at the installation site. The construction of the noise barriers at the installation site allows greater flexibility and better adaptation to the terrain on which the noise barriers are placed, while the advantage of blocks or modules manufactured in advance in a controlled environment is greater uniformity, better durability and regularly lower costs. A disadvantage of such blocks or modules is the need to allow access and space to work machines (e.g. cranes and transport vehicles) at the installation site area in order to enable the assembling of the whole noise barrier.

2.2.3 Free standing noise barriers

Free standing noise barriers can be concrete noise barriers moulded at the installation site. The process of building them involves digging the ground for support, laying down a steel reinforcement, pouring concrete, surface treatment and concrete hardening. In this particular construction, the casting and later hardening of concrete are carried out in different weather conditions, which can affect the quality of the final product. The advantage of these noise barriers is the fact that the shape and method of installation can be fully adapted to the terrain, which is the reason why these noise barriers are most commonly used on bridges and viaducts. An additional advantage of such noise barriers is their high structural strength and resistance to damage, which is why, alongside the noise protection and reduction function, they are very often used as retaining walls for separating traffic lanes for safety reasons.

Free standing noise barriers can also be precast or premanufactured, i.e. concrete panels are factory-made under controlled conditions and then delivered to the installation site where they are installed (please see **Figure 6**). Furthermore, free standing noise barriers can be designed as green vertical noise barriers which are shown in **Figure 7**. Currently green vertical noise barriers are becoming more and more popular in cities because, in addition to reducing noise, they also reduce air pollution and they do not take up additional space, i.e. they are built into existing freestanding walls or facades.

To conclude, all of the aforementioned types of noise barriers are used as a measure of protection against traffic noise, while the choice of the noise barrier itself depends on the noise level at the location where the noise barrier is installed (i.e. acoustic properties of the sound source). In addition, the selection depends on the position of the noise barrier itself (distance of the noise barrier from the sound source and the receiver) and the allowed noise barrier height.

2.3 Material features

The noise barrier construction can be made of different materials. It is possible to construct a noise barrier with only one material; however, more often the construction of noise barriers consists of several different materials. The choice of materials depends on several basic factors: acoustic properties, type and level of noise sources from which we are protecting a certain space, mechanical properties, aesthetic requirements on both sides of the noise barrier, regulations and the cost of an investment in noise protection for a certain space. In addition to the above-mentioned and described basic materials (concrete, metals, wood, etc.), sound-absorbing materials (e.g. stone wool) are often used in practical case scenarios. Such materials can be used as noise barriers' fill and with their sound absorption properties increase the noise barrier efficiency.

Figure 7.
An example of a transparent noise barrier on expressway.

2.3.1 Concrete

Concrete is one of the most commonly used building materials. Concrete cast into blocks which are transported to the installation site or cast at the installation site is considered as one of the most durable construction. It is robust and can withstand high temperatures, strong sunlight, moisture, ice and salt. It is quite easy to shape and colour; thus, its appearance can vary. The versatility of concrete also relates to the shape and size of the slabs that can be produced (cast in place, prefabricated concrete blocks). In addition, concrete enables various installation techniques.

2.3.2 Metals

Three types of metal are most commonly used while constructing a noise barrier: steel, aluminium and stainless steel. Steel is the cheapest and most common of all metals used generally in construction. Thus, it is also generally used in the noise barrier construction, especially combined with concrete. Steel consists of a mixture of iron ore, coal and a small amount of other metals, while the ratio of the constituents varies depending on the desired physical properties.

For structures that require a slightly lower mass, aluminium is used, mainly as a light alloy with additives of manganese, silicon, copper and/or magnesium. Depending on the type of elements added to the aluminium in the alloy and their ratio, different mechanical, thermal, industrial and acoustic properties are obtained. Aluminium and its alloys are weatherproof and can be easily coated and anodized in different colours, making them suitable for installations with specific aesthetic requirements.

Stainless steel, which is a low carbon alloy with a minimum of 10.5% chromium, and is often mixed with nickel, molybdenum and titanium, is a very durable and resistant to corrosion due to its chromium alloy's ability to bind to oxygen atoms from the air, thus creating an invisible thin protective film on the metal surface that protects the metal from oxidation and damage. Since stainless steel is almost completely resistant to corrosion, its surface does not need to be coated or additionally protected and is often used in areas with high humidity, especially if the noise barrier is in contact with or near seawater.

Metal panels have a great advantage over concrete materials, which is their light weight. Their low weight makes them particularly useful for vertical extensions of existing walls, that is, for installation on existing retaining walls and for installation on bridges. Due to their simple manufacturing and easy assembly, either by attachment or welding, they are often used on bridges and viaducts attached to the existing load-bearing structural elements of the bridge itself.

2.3.3 Wood

Different types of wood can be used in the production of noise barriers. The design range of constructions varies from simple, i.e. consisting of several wooden panels, to very complex structures made of multiple wooden pieces that can often be made of different types of wood. Wood is a natural, environmentally friendly material, which is very easy to process and has a low mass. Panels of wood that are creating a noise barrier parts can be installed piece by piece at the installation site or may be partially assembled on the ground prior to installation. Such noise barriers are easy to disassemble or remove, and the wood is from the aesthetic point of view, a very accepting and pleasing material to the environment. In addition, it does not conduct electricity. A significant problem with noise barriers made of wood is its flammability, and furthermore the smoke and gases resulting from its combustion are toxic. In addition to burning, the process of wood rotting in contact with moisture is very fast, which makes it necessary to protect it with a chemical preservative, which adds complexity to the process of producing the noise barriers and the need for more frequent maintenance. Wood products are not dimensionally stable and tend to change shape, which causes open cracks between joints, and the tendency to change shape increases with the dimensions of the wooden piece of the noise barrier itself.

2.3.4 Transparent panels

Transparent panels can be made of glass or plastic materials, such as Plexiglas, Lexan, Acrylic, etc. which is shown in **Figure 7**.

Glass panels are usually made of tempered or laminated tempered glass. Tempering the glass strengthens the glass, and therefore such product becomes more resistant to breakage. If broken, the shards are small and grainy, with pieces generally no larger than 12 mm, which is far safer than knife-like shards that result from breaking glass that has not been heat treated. In addition to tempering, the glass panel can also be laminated. This type of glass is manufactured by inserting a translucent, rubber and flexible interlayer between the two tempered glass panels.

When such glass breaks, small granular fragments are formed which remain glued to the interlayer.

Transparent panels are ideal for reducing or completely eliminating the visual impact of a noise barrier; however, their costs can be 20 times higher than those made of concrete or steel. The justification for their high cost can be found in improving safety in places where opaque noise barriers can have a negative effect on visibility. These types of panels are more sensitive to damage from flying debris and abrasive action as a consequence of the sandblasting effect that is inevitably due to the swirling dust that is always present on the pavement layer.

2.3.5 Plastic

There are several types of plastic materials available and often used in the construction of noise barriers: polyethylene, PVC (polyvinyl chloride) and fibre-glass. Plastic panels can be installed in almost any situation due to their extremely low mass, easiness to mould and weatherproof features. Bearing in mind all of the aforementioned, they are increasingly used for the construction of noise barriers, especially those of a more complex shape. The problem with plastic materials is a slightly lower structural strength and flammability, i.e. the smoke and gases produced by the combustion of plastics are very toxic.

2.3.6 Composite materials

Composite materials for noise barriers can be defined as any product composed of two or more "basic" materials, for example, wood mixed with concrete and then placed on a concrete foundation. By combining basic materials, the characteristics of the final product (noise barriers) and its durability, and even in some cases safety, are altered.

Material	Thickness (mm)	Weight (kg/m2)	Transmission loss (dBA)
Concrete Block, 200 mm × 200 mm × 405 mm	200	151	34
Dense Concrete	100	244	40
Steel	1,27	10	25
Steel	0,95	7,3	22
Steel	0,79	6,1	20
Iron	0,64	4,9	18
Aluminium	1,59	4,4	23
Aluminium	3,18	8,8	25
Aluminium	6,35	17,1	27
Wood (Fir)	12	8,3	18
Wood (Fir)	25	16,1	21
Wood (Fir)	50	32,7	24
Plywood	12	8,3	20
Plywood	25	16,1	23
Glass	3,18	7,8	22
Plexiglas	6	7,3	22

Table 1.
Approximate values of transmission loss parameter for different types of materials.

2.4 Noise barriers transmission loss (*TL*)

Typical values of the noise barriers transmission loss (TL) parameter when looking at the A-weighted characteristic are from 10 dBA to 15 dBA. Noise barriers should be constructed of materials with a minimum density of 20 kg m^2. A density of 20 kg/m^2 can be achieved by lighter and thicker or heavier and thinner materials (i.e. higher material density enables a thinner material). **Table 1** gives the approximate TL values for some common materials, tested for a typical A-weighted highway traffic frequency spectrum [16]. These values can be used as a rough guideline in designing noise barriers. For more accurate values, one would need to find material testing reports from authorised laboratories.

3. A contemporary approach: soundscape

The soundscape concept has been introduced to modify and complement the assessment of noise and its effects on humans. From the beginning, soundscape research has questioned the limits of existing acoustic measurements and the cultural dimension which Schafer included in his research for the first time in the 1960s [17].

3.1 Soundscape definition

A particular soundscape includes all the sounds from a certain acoustic environment received by the human ear. The first idea of soundscape was introduced by Schafer, in his book *The New Soundscape* [17]. His primary idea was to record a soundscape of the world in a form of a map similar to geographical maps. However, extremely rapid changes in the soundscapes have made this idea impossible to implement. Soundscapes, among other things, change rapidly due to the growth of the human population, people's migration and traffic increase. On the other hand, it is possible and of great interest to record current soundscapes.

The soundscape of a certain environment consists of various sound groups and sound sources. They can be divided into three major groups, biophony, geophony and anthrophony [17], which is shown in **Figure 8**.

Biophony are all the sounds produced by living organisms in their natural habitat (**Table 2**). It is by far the most complex feature of soundscape because it combines all biological sound sources, from microscopic to large fauna that live in a given environment for a certain period of time. In environments that are rich with different voices of living beings, organisms produce acoustic signals in different spatial relationships which can sound as one or more sound signals. Geophony are all-natural sounds coming from non-biological sources in a certain environment (**Table 2**). Generally, they can be divided into four types: the sound effects of wind, water, climate and geophysical forces. Anthrophony are all sounds generated by humans in any natural environment. This group includes sounds coming from people, music and traffic noise.

Bearing in mind all of the aforementioned, it can be concluded that the concept of soundscape as a field of research is extremely broad and requires a multidisciplinary approach. In studies and researches, apart from acoustic engineers, psychologists, physicians, builders, architects and sociologists should be involved.

Table 2 shows sound sources or acoustic components and direct acoustic effects of non-anthropogenic sound elements (biophony and geophony).

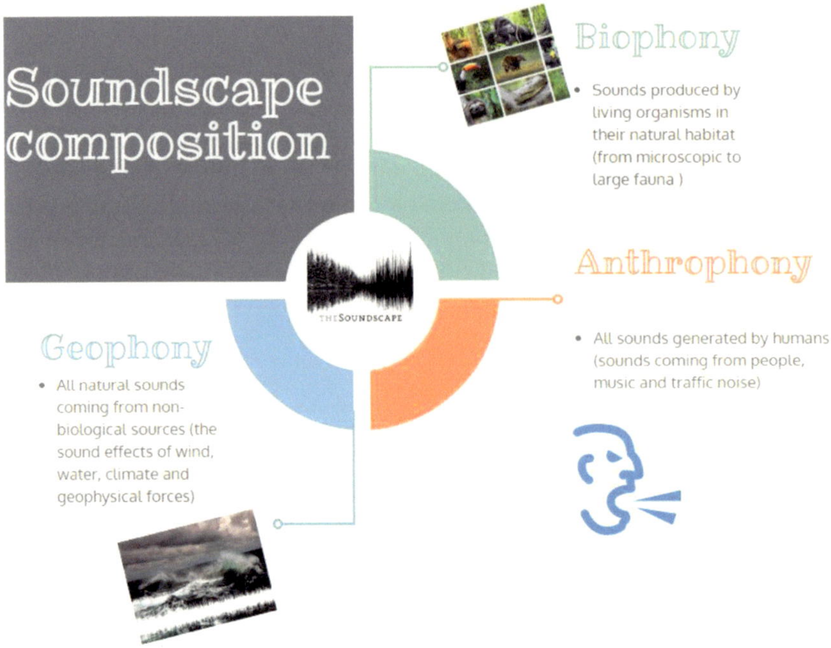

Figure 8.
Soundscape composition.

Weather
RS: wind, rain, thunder, earthquake, clouds
D: sound propagation, absorption, reflection, refraction, diffusion, diffraction, masking
Animals
S: sound of animals: birds, mammals, insects, reptiles, amphibians; moving of animals: running, flying, jumping, landing; non-voice expressions: moving of insects' wings
D: sound propagation, masking
Nature
S: rivers and streams, waves, flux and reflux, vegetation excitation e.g. leaf rustle, falling of trees
D: propagation of sound, masking, absorption, reflection, refraction, diffusion, diffraction
Terrain
S: barriers, dams
D: propagation of sound, absorption, reflection, refraction, diffusion, diffraction
S—Sound sources or acoustic components and D—direct acoustic effects.

Table 2.
Characteristics of non-anthropogenic sound elements (based on [17])

3.2 Soundscape classification

The most common soundscape classification is the one with respect to the related environment, and therefore we can distinguish them as follows (**Figures 9** and **10**):

a. Natural soundscapes (e.g. marine, forest soundscape, etc.)

b. Rural soundscapes.

c. Urban soundscapes.

Figure 9.
Soundscape classification.

Figure 10.
Description of soundwalking (based on [18]).

Considering the way and style of today's life, it can be concluded that the urban soundscape is most explored and is changing in a fast pace. A city's soundscape encompasses all three active components which describe a certain soundscape; however, the largest impact is anthropogenic, i.e. sounds generated by various human activities. Looking through history, after the industrial and electrical revolution, the look and sound of the city have changed remarkably. Today a similar thing is happening, however as a result of the accelerated construction work and overcrowding of cities. One of the biggest problems in the city has become noise, and the biggest source of this noise is traffic [19].

3.3 How to record a soundscape?

One of the possible ways to record a soundscape is the soundwalk method. *Soundwalk* method, as a concept, was first introduced by an urban planner Kevin

Lynch [20, 21]. His idea was to follow the usual routes which people use on their commute and which are specific to places of interest with the goal of "capturing" their soundscape. The usual recording of a soundscape has the duration of 30 minutes. The 30-minute choice corresponds to the distance a man would walk across in an average European city and, on the other hand, keeps a certain homogeneity as far as activities in that particular soundscape are concerned. Recording takes place several times a day, for several days, however always at a nice and dry weather. Depending on the research premises, the question how to exactly and accurately process the data obtained from the recordings remains. Although most scientists in their studies claim that the *soundwalk* "walks" lasted for an average of 15 minutes, it is possible to record soundscapes for a longer period depending on the focus and goal of the research. It is also possible to "cut" or shorten the longer soundscape recordings and apply other acoustic "tools" to the soundscape recordings.

The *soundwalk* method uses a recorder and a pair of binaural microphones places in the ears of the person who is performing the *soundwalks*, i.e. *soundwalker*. The *soundwalker*, as the name suggests, performs "walks" through a certain environment. The *soundwalker* needs to carefully breath while walking, and this kind of recording is performed normally in dry and sunny weather, unless we want to record a certain natural manifestation, such as rain. Recording is performed at the height of the walker (the binaural microphones are placed in the ears of the walker) and therefore the recorded signals are similar (at the highest possible level) to the signals heard by the pedestrians in that environment (**Figure 11**).

There is still no consensus among scientists regarding the recording length of the soundscape. Namely, there is no clearly defined time limit that would determine the difference between authentically play backed soundscapes while simultaneously avoiding the listener's fatigue. Reproduction of soundscapes in terms of its length depends on the researcher and the purpose of his research from a few minutes to several hours.

3.4 Soundscape analysis

Alongside the soundwalk method, the researchers usually perform the analysis of the recorded soundscape with the "help" of the questionnaire which is again not defined by any standard or norm. The concept of the questionnaire, whose purpose is a detailed analysis provided by the listener, is not clearly defined, and moreover it depends again on research premises and focus [22–24]. Questionnaires vary from direct questions to listeners about the soundscape, requirements for a more detailed descriptions of the soundscape in terms of defining them as pleasant or unpleasant, attributes that may or may not be related to mathematical scales and adjective pairs that are not standardised so each researcher can use "their own" adjective pairs which they consider to best describe the soundscape and fit their research. An example of questionnaire [25] which uses bipolar adjective pairs is shown in **Table 3**.

Figure 12 shows an example of performed soundscape analysis, in particular an expressway which stretches from the east to the west exit of the Zagreb, capital of Croatia [26]. To be more precise, **Figure 12** shows a photograph of the recorded location using the *soundwalking* method, the route of the *soundwalker*, the spectrogram of the recording and the sound sources which were characterised as distracting obtained through a specially designed questionnaire for that particular research [26].

To sum it up, the questionnaires are not an objective acoustic parameter; however, they present a good measure of the listener's perception of the soundscape. A major problem is how to correlate such a subjective parameter with objective acoustic parameters, i.e. loudness, sharpness, roughness and fluctuation strength [27].

Figure 11.
An example of the soundwalker's route.

Pair no.	Assigned bipolar adjective pair	Type of pair
1	quiet – loud	Auditory
2	deep - high-pitched	
3	diverse - monotonous	
4	pleasant - unpleasant	Emotional
5	dynamic – calm	
6	natural - artificial	
7	harmonious - chaotic	
8	appealing - repulsive	
9	soothing - stressful	
10	conspicuous – inconspicuous	
11	gentle – rough	

Table 3.
The adjective pairs used in questionnaire [25].

3.5 Using the soundscape as a noise reduction instrument

The ISO 12913-1 standard defines soundscapes as acoustic environments "as perceived by people, in context". Thus, nowadays more and more soundscape studies are oriented towards human health, well-being and overall quality of life [28–31]. In addition, the WHO Environmental Noise Guidelines for the European Region provide certain guidance on protecting human health from harmful exposure to environmental noise. The guidelines strongly recommend reducing the noise levels (L_{den} and L_{night}) for the cases of environmental noise sources such as road traffic noise, railway noise, aircraft noise, wind turbine noise and leisure noise [32]. In the past the only possible approach to resolve this issue and reduce noise levels in an efficient way has been noise barriers which have been described thoroughly in previous sections. However, in order to construct and position an effective noise barrier, first one needs to have enough space for it, i.e. noise barriers can serve as a solutions only if they are planned before the actual building which is today a quite rare case scenario. In addition, if there is an opportunity to incorporate a noise barrier into an existing urban environment, researchers should take into account the "visual pleasantness" as well as the economic feasibility of the noise barrier.

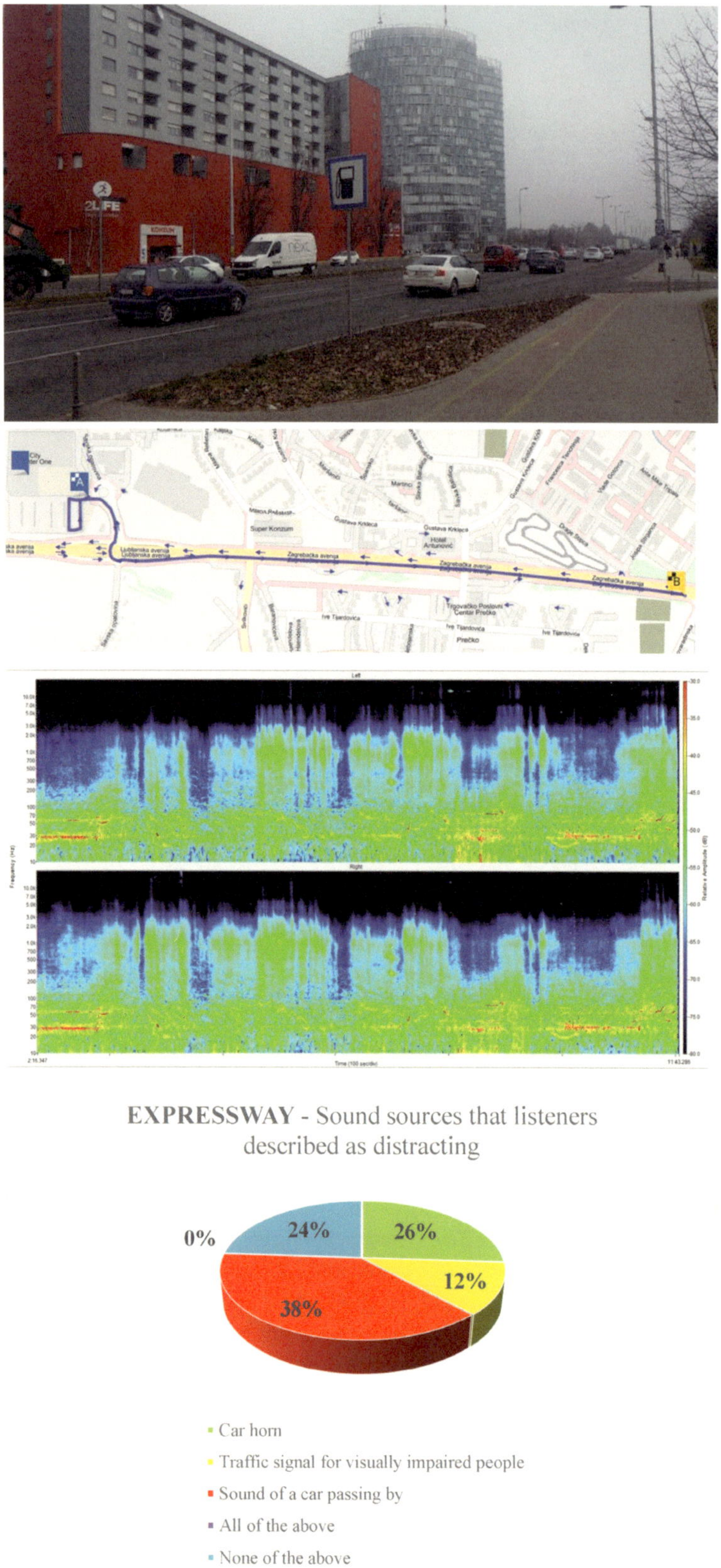

Figure 12.
Soundscape analysis.

Finally the noise barriers as a solution are focused only on reducing the noise levels from traffic sources, while the common and final goal of the WHO and researchers, with different types of expertise, is to improve the overall quality of life by using different tools, guidelines, descriptors and an interdisciplinary approach by designing, preserving and investigating pleasant acoustic environments, i.e. positive soundscapes [33–36].

There are several paths where soundscape research can be used as an instrument for reducing noise pollution.

One of them is discovering what in particular makes a certain acoustic environment perceived as pleasant and then utilising those sound sources in negatively perceived soundscapes. Several soundscape studies have shown that the majority of population responds very well to the sound of water and birds singing, and therefore, those sound signals can be used for acoustic masking of unpleasant parts of other acoustic environments which may be rated as less pleasant [37].

When considering negatively perceived indoor soundscape, improvements regarding noise reduction can actually be achieved using specific acoustic absorptive materials and collaborating with architects regarding the layout of, e.g. open-space offices [38].

Another solution can be mixing soundscapes with music. Music is something that is deeply personal, something that consumes us, and it alters our mood. It can lift us up, it can comfort us, and it can make us feel calm and relaxed. Because of that recent studies show good results in creating innovative "music soundscapes" which can be a powerful tool for healing the people from a stressful and overvibrant urban everyday life. To be precise by installing special gazeboes in urban parks and providing the user to combine the preferred music with an existing soundscape, one can create calming and relaxing zones that could provide a short music break from everyday obligations and help to endure the day in a better mood [39].

4. Combining the traditional and contemporary approach

When considering only noise barriers as an instrument to combat noise and noise pollution, they have the ability to reduce noise levels by 3–20 dB. Over 3000 km of noise barriers have been installed alongside European rail networks. They are even more widely used alongside roads, including countries such as Austria, Denmark, France, Germany, Italy, Poland, Spain and Netherlands. Keeping that in mind, it can be concluded that noise barriers as a solution are mainly focused on reducing the noise levels from traffic sources, while the goal of the WHO and many researchers is to improve the overall quality of life by using different tools, guidelines, descriptors and an interdisciplinary approach by designing, preserving and investigating positively perceived soundscapes as previously mentioned.

For each noise barrier, its acoustic performance can be determined, as described in Section 2.1. Noise levels reduction with noise barriers. Here are some practical examples regarding the different performance of several noise barriers:

- First, the results obtained by simulation for five simple 5-m straight noise barriers (made of different materials) are presented. **Figure 13** shows the parameter of average noise reduction (*ILavg*) depending on the receiving position [40].

- In [41] insertion loss values at 1/1 octave band were calculated bearing in mind the noise barrier types, while the height of all noise barrier was fixed at 5 m and the receiver's position was 2 m from the noise barrier. Prediction of sound pressure levels for five types of road traffic noise sources attenuated by different types of barriers was conducted using acoustic simulation software Enpro (Environment Noise Prediction and Design Program). Predicted insertion loss data is shown **Table 4**.

- In [42] a comparison of the obtained values for the green wall sound absorption coefficient and different common building materials was carried out (**Figure 14**).

In addition, for each type of noise barrier, its visual stimulus can be determined by using soundscape research which is described in [41] where images of five types of noise barriers were investigated. In the aforementioned research, timber, metal, transparent glass, vegetation and concrete barriers were created using Adobe Photoshop CS4 software. A viewpoint was fixed in order to avoid influence of view angles. Furthermore, small and large portion of ivy images were covered on the transparent and concrete barriers images in order to evaluate the visual effect of vegetation.

Furthermore, in [43] a case study is presented, where a sample of residents living close to a railway line assessed noise-related aspects for several barriers with different visual characteristics in an IVR laboratory test. In particular, three main factors were analysed: the barrier type concerning the visibility of the noise source through the screen, the visual aspect of the barrier concerning some aesthetic

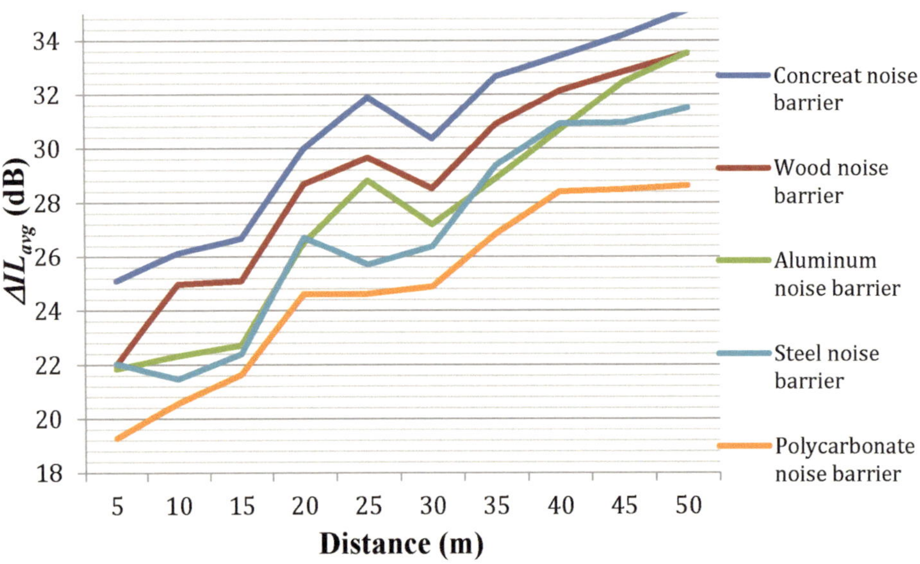

Figure 13.
Parameter of average noise reduction (ILavg) depending on the receiving position [40].

Frequency (Hz)	63	125	250	500	1 k	2 k	4 k	8 k
Timber	2.0	2.5	9.7	13.1	15.9	17.9	18.0	18.0
Metal	2.2	7.7	16.6	19.0	19.6	18.9	19.0	18.7
Transparent	2.0	3.4	7.8	12.7	14.7	14.9	15.4	15.7
Concrete	14.7	14.4	17.7	19.0	19.6	19.9	20.0	20.0
Vegetation	6.3	7.3	12.3	17.0	18.6	18.9	19.3	18.0

Table 4.
Predicted insertion loss data [41].

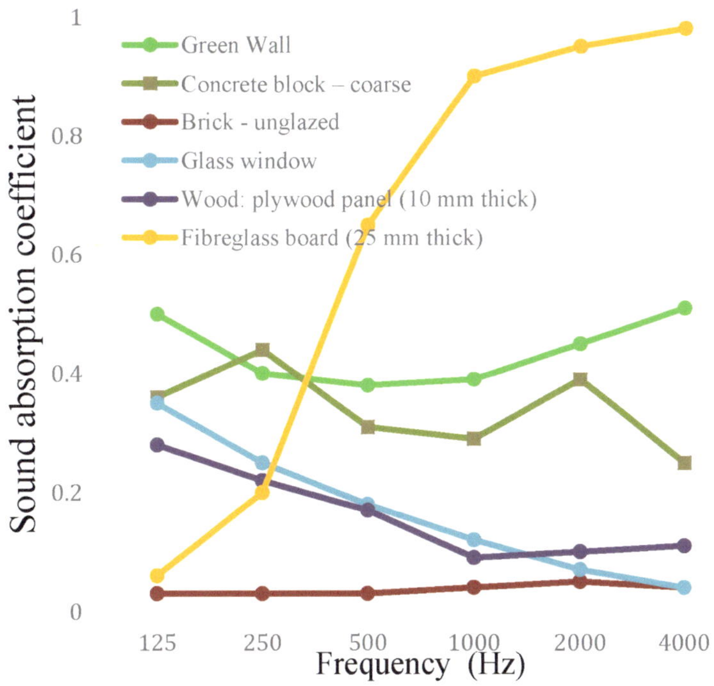

Figure 14.
Comparison of sound absorption coefficient for the green wall and common building materials.

issues and the noise level at the receiver concerning the acoustic performance of the barrier and the magnitude of the sound source. The main results of the ANOVA analysis showed that for transparent barriers, perceived loudness and noise annoyance were judged lower than for opaque barriers; this difference increased as noise level increased.

A much more effective way of reducing noise and noise pollution is to use noise barriers and soundscape together, by incorporating noise barriers, auditory ratings and visual assessment. Here we note that this way allows the design of better noise barriers and soundscapes and thus better acoustic urban environments. It has been proven that a noise barrier that better attenuates low frequencies such as a concrete noise barrier has a better acoustic rating. When considering user ratings, it has been proven in [41] that people respond much better to noise barriers made from natural materials and especially green walls, i.e. the visual pleasantness is much higher for that type of noise barriers.

To sum it up, the best way to tackle this burning issue, i.e. noise pollution, is the combination of both methods (noise barriers and soundscape research). Both methods complement each other in a very good and effective way while moving and in a way breaking the limitations which each individual method has.

5. Conclusions

In this chapter two different approaches for noise reduction have been described and discussed in detail. Finally, by comparing these approaches, it can be concluded that each one of them has its advantages, disadvantages and limitations. The final choice of noise reduction measure in most cases depends on the limitations regarding the location, cost, etc. It can be concluded that the best results regarding the reduction of noise pollution can be obtained by combining both described approaches. The approaches work very well and effective together and, in that way, extend the limitations which occur when using only one approach.

Bearing in mind all of the aforementioned, urban planners, architects, doctors, psychologists as well as acoustic engineers should work together and benefit from each other's work with a common cause to improve the overall quality of life. By collaborating together, it is possible to reduce noise pollution and moreover improve the human health and well-being of the residents, especially the ones living and working in urban areas.

Author details

Mia Suhanek* and Sanja Grubesa
Department of Electroacoustics, Faculty of Electrical Engineering and Computing, University of Zagreb, Zagreb, Croatia

*Address all correspondence to: mia.suhanek@fer.hr

References

[1] Koussa F, Defrance J, Jean P, Blanc-Benon P. Acoustical efficiency of a sonic crystal assisted noise barrier. Acta Acustica United with Acustica. 2017;**99**(3):399-409. DOI: 10.3813/AAA.918621

[2] Grubeša S, Domitrović H, Jambrošić K. Performance of traffic noise barriers with varying cross-section. Promet: Traffic & Transportation. 2011;**23**:161-168. DOI: 10.7307/ptt.v23i3.119

[3] Fard Samaneh M, Peters H, Marburg S, Kessissoglou N. Acoustic performance of a barrier embedded with Helmholtz resonators using a quasi-periodic boundary element technique. Acta Acustica United with Acustica. 2017;**103**(3):444-450. DOI: 10.3813/AAA.919074

[4] Ishizuka T, Fujiwara F. Performance of noise barriers with various edge shapes and acoustical conditions. Applied Acoustics. 2004;**65**:125-141. DOI: 10.1016/j.apacoust.2003.08.006

[5] Baulac M, Defrance J, Philippe J. Optimisation with genetic algorithm of the acoustic performance of T-shaped noise barriers with a reactive top surface. Applied Acoustics. 2008;**69**(4):332-342. DOI: 10.1016/j.apacoust.2006.11.002

[6] Monazzam MR, Lam YW. Performance of T-shape barriers with top surface covered with absorptive quadratic residue diffusers. Applied Acoustics. 2008;**69**(2):93-109. DOI: 10.1016/j.apacoust.2006.10.006

[7] Greiner D, Aznárez JJ, Maeso O, Winter G. Single- and multi-objective shape design of Y-noise barriers using evolutionary computation and boundary elements. Advances in Engineering Software. 2010;**41**(2):368-378. DOI: 10.1016/j.advengsoft.2009.06.007

[8] Toledo R, Aznárez JJ, Maeso O, Greiner D. A procedure for improving the acoustic efficiency of top-edge devices on noise barriers: An application of genetic algorithms and boundary elements. In: Evolutionary Algorithms and Metaheuristics in Civil Engineering and Construction Management. Barcelona, Spain: Springer; 2015. DOI: 10.1007/978-3-319-20406-2_7

[9] Toledo R, Aznárez JJ, Greiner D, Maeso O. Shape design optimization of road acoustic barriers featuring top-edge devices by using genetic algorithms and boundary elements. Engineering Analysis with Boundary Elements. 2016;**63**:49-60. DOI: 10.1016/j.enganabound.2015.10.011

[10] Maffei L, Masullo M, Aletta F. Influence of the design of railway noise barriers on soundscape perception. In: Internoise 2012; New York. 2012. DOI: 10.13140/2.1.4899.9369

[11] Maffei L, Masullo M, Aletta F, Gabriele M. The influence of visual characteristics of barriers on railway noise perception. Science of The Total Environment. 2013;**445-446**:41-47. DOI: 10.1016/j.scitotenv.2012.12.025

[12] Grubeša S, Jambrošić K, Domitrović H. Noise barriers with varying cross-section optimized by genetic algorithms. Applied Acoustics. 2012;**73**(11):1129-1137. DOI: 10.1016/j.apacoust.2012.05.005

[13] Grubeša S, Suhanek M, Đurek I, Petošić A. Combined acoustical and economical noise barrier optimization using genetic algorithms. GRAĐEVINAR. 2019;**71**(3):177-185. DOI: 10.14256/JCE.2523.2018

[14] EN 1793-1-2017. Road traffic noise reducing devices—Test method for determining the acoustic performance—Part 1: Intrinsic

characteristics of sound absorption under diffuse sound field conditions

[15] EN 1793-2-2017. Road traffic noise reducing devices—Test method for determining the acoustic performance—Part 2: Intrinsic characteristics of airborne sound insulation under diffuse sound field conditions

[16] Gregg G, Fleming G, Knauer HS, Lee CSY, Pedersen S. Noise Barrier Design Handbook. Available from: https://www.fhwa.dot.gov/environment/noise/noise_barriers/design_construction/design/index.cfm. [Accessed: 13 January 2019]

[17] Schafer RM. Our Sonic Environment and the Soundscape: The Tuning of the World. Rochester, Vermont: Destiny Books. 1994. pp. 256

[18] Westerkamp H. Inside the Soundscape. Available from: https://www.hildegardwesterkamp.ca/sound/installations/Nada/soundwalk/, [Accessed: 10 January 2020]

[19] Fiebig A, Guidati S, Goehrke A. Psychoacoustic Evaluation of Traffic Noise. In: NAG/DAGA 2009. Rotterdam; 2009

[20] Lynch K. The Image of the City. Cambridge, Massachusetts, London, England: The MIT Press; 1960

[21] Semidor C. Listening to a city with the soundwalk method. Acta Acustica United with Acustica. 2006;**92**:959-964

[22] Schulte-Fortkamp B. The Concept of Meaning in Soundscapes with Respect to the New Experts. In: COST. March 2010

[23] Genuit K, Bray W. Soundscape measurement and analysis. Journal of the Acoustical Society of America. 2006;**119**:3260. DOI: 10.1121/1.4786096

[24] Kull RC. Soundscape measurements: Moving towards a standard. Journal of the Acoustical Society of America. 2005;**116**:1875. DOI: 10.1121/1.4779307

[25] Suhanek M, Đurek I. Implementation of bipolar adjective pairs in analysis of urban acoustic environments. Promet-Traffic & Transportation. 2016:461-470. DOI: 10.7307/ptt.v28i5.2089

[26] Suhanek M. Evaluation of soundscapes regarding sudden and unexpected sound changes [thesis]. Zagreb: Faculty of EE and Computing, University of Zagreb; 2013

[27] Zwicker E, Fastl H. Psychoacoustics: Facts and Models. Berlin, Heidelberg, New York: Springer-Verlag; 1999. pp. 220

[28] Schulte-Fortkamp B, Fiebig A. Soundscape analysis in a residental area: An evaluation of noise and people's mind. Acta Acustica United with Acustica. 2006;**92**:875-880. DOI: 10.1121/1.3298437

[29] Schulte-Fortkamp B, Kang J. Introduction to the special issue on soundscapes. Journal of the Acoustical Society of America. 2013;**134**:765-766. DOI: 10.1121/1.4810760

[30] Botteldooren D, Andringa T, Aspuru I, Brown AL, Dubois D, Guastavino C, et al. From sonic environment to soundscape. Soundscape and the Built Environment. Chapter 2. 2015:1-26. DOI: 10.1201/b19145-3

[31] Kang J, Aletta F, Gjestland TT, Brown LA, Botteldooren D, Schulte-Fortkamp B, et al. Ten questions on the soundscapes of the built environment. Building and Environment. 2016;**108**:284-294. DOI: 10.1016/j.buildenv.2016.08.011

[32] Environmental Noise Guidelines for the European Region. Europe UN City, Denmark: WHO Regional Office; 2018

[33] Lercher P, Van Kamp I, Von Lindern E, Botteldooren D. Perceived soundscapes and health-related quality of life, context, restoration, and personal characteristics. Soundscape and the Built Environment. Chapter 5. 2015:90-125. DOI: 10.1201/b19145-6

[34] Cerwén G. Urban soundscapes: A quasi-experiment in landscape architecture. Landscape Research. 2016;**41**:481-494. DOI: 10.1080/01426397.2015.1117062

[35] Xiao J, Lavia L, Kang J. Towards an agile participatory urban soundscape planning framework. Journal of Environmental Planning and Management. 2018;**61**:677-698. DOI: 10.1080/09640568.2017.1331843

[36] Bild E, Pfeffer K, Coler M, Rubin O, Bertolini L. Public space users' soundscape evaluations in relation to their activities. An Amsterdam-based study. Frontiers in Psychology. 2018;**9**:1593. DOI: 10.3389/fpsyg.2018.01593

[37] Lercher P, Schulte-Fortkamp B. The relevance of soundscape research to the assessment of noise annoyance at the community level. In: Proceedings of the 8th International Congress on Noise as a Public Health Problem; Rotterdam. 2003

[38] Thomson EA. The soundscape of modernity: Architectural acoustics and the culture of listening in America. Books.google.com; 2004. DOI: 10.1017/S0007087403315390

[39] Steele D, Bild E, Tarlao C, Guastavino C. Soundtracking the public space: Outcomes of the musikiosk soundscape intervention. International Journal of Environmental Research and Public Health. 2019;**16**:1-38. DOI: 10.3390/ijerph16101865

[40] Grubeša S. Optimization of noise barriers parameters using numerical procedures [thesis]. Zagreb: Faculty of EE and Computing, University of Zagreb; 2011

[41] Hong JY, Jang HS, Jeon JY. Evaluation of noise barriers for soundscape perception through laboratory experiments. In: Proceedings of the of the Acoustics 2012 Nantes Conference, 2012; Nantes

[42] Azkorra Z, Pérez G, Coma J, Cabeza LF, Bures S, Álvaro JE, et al. Evaluation of green walls as a passive acoustic insulation system for buildings. Applied Acoustics. 2015;**89**:46-56. DOI: 10.1016/j.apacoust.2014.09.010

[43] Maffei L, Masullo M, Aletta F, Gabriele M. The influence of visual characteristics of barriers on railway noise perception. Science of the Total Environment. 2013;**445-226**:41-47. DOI: 10.1016/j.scitotenv.2012.12.025

Chapter 5

Traffic Noise

Sanja Grubesa and Mia Suhanek

Abstract

Currently, noise pollution is a major problem especially in urban areas, and moreover traffic noise is the most significant source of noise in cities. A large number of cars and other road vehicles that have internal combustion engines are making road traffic noise a leading noise pollution source. Electric and hybrid cars, which are nowadays slowly replacing them, give rise to lower noise level in urban areas as their engines are generally silent. However, the mere absence of internal combustion engines cannot be the only measure for lowering noise levels in urban areas. The goal of this chapter is to define and describe traffic noise, the reasons for its occurrence, and all existing ways of reducing traffic noise.

Keywords: noise, noise pollution, traffic noise, electric and hybrid cars, noise reduction

1. Introduction

Nowadays noise pollution is the focus of various studies and research due to its proven significant impact on human health and work efficiency. Research shows that traffic noise in urban areas has tremendously increased since the beginning of the century, primarily due to increased transportation of people and goods. It can be concluded that in urban areas the largest source of noise is traffic-induced noise, which accounts for 80% of all communal noise sources. Traffic noise caused by road traffic is the most common type of noise in urban areas and as such poses a serious problem. **Figure 1** shows the distribution of human noise annoyance according to the type of noise source [1].

According to **Table 1**, provided by the International Union of Railways (UIC), all types of trains produce less noise than trucks, cars, airplanes, and other means of transport. Railway is the most favorable form of transport, in terms of noise as an influential factor for environmental degradation and human health. Therefore, it can be determined that the railway has the lowest share of noise in urban areas among other means of transport.

2. Traffic noise sources

2.1 Road traffic noise

Road traffic noise depends on the following three factors:

- Type of road vehicles.

- Friction between the vehicle wheels and the road surface.

- Driving style and driver behavior.

When considering vehicles that have an internal combustion engine (ICE) as the noise source, most of the noise comes from the sources or systems shown in **Figure 2**. The aforementioned sources and systems are explained in detail in the following paragraph.

Engine noise is created during the process of compression and expansion in the engine, which creates engine vibrations which then emit noise. The engine noise depends on the engine volume, speed, and capacity. The suction system noise is caused by the opening and closing of the suction valves, and furthermore the intensity of such noise depends on the mode of operation of the engine, the speed of the engine itself, and the type of air filter. Noise from the exhaust system is created by the sudden release of gas into the exhaust system itself in order to open the exhaust valve. The fan noise is generated due to the operation of the fans in the vehicle, and the fans generally produce a broadband noise. Tire noise occurs when the tires and

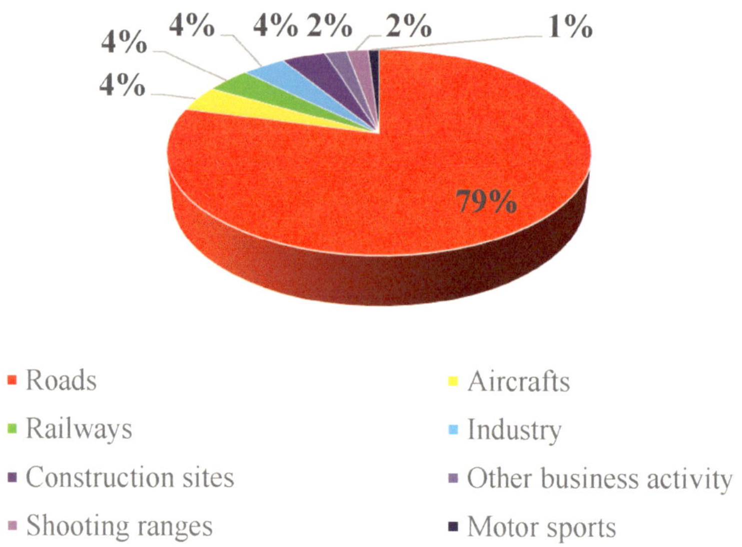

Figure 1.
Distribution of human noise annoyance according to the type of noise source [1].

Type of vehicle	Average noise level [dBA]
Car (700–1300 cm^3)	82
Motorcycle	90
Heavy cargo truck	103
Turbojet airplane	150
Fast passenger train	65
Cargo train (speed up to 120 km/h)	60
Local train	70

Table 1.
The average noise level generated by different types of vehicles (International Union of Railways (UIC)).

road surfaces come into contact. This type of noise depends on the type of road surface, the tire construction, and finally the speed and driving style [2].

In terms of noise pollution, electric vehicles represent the future, especially when compared to vehicles with an internal combustion engine (see **Figure 3**). However, at low speeds, electric vehicles produce very small levels of noise, i.e., in current acoustic urban environments, they are practically inaudible. For example, the noise level difference between an electric vehicle and an internal combustion

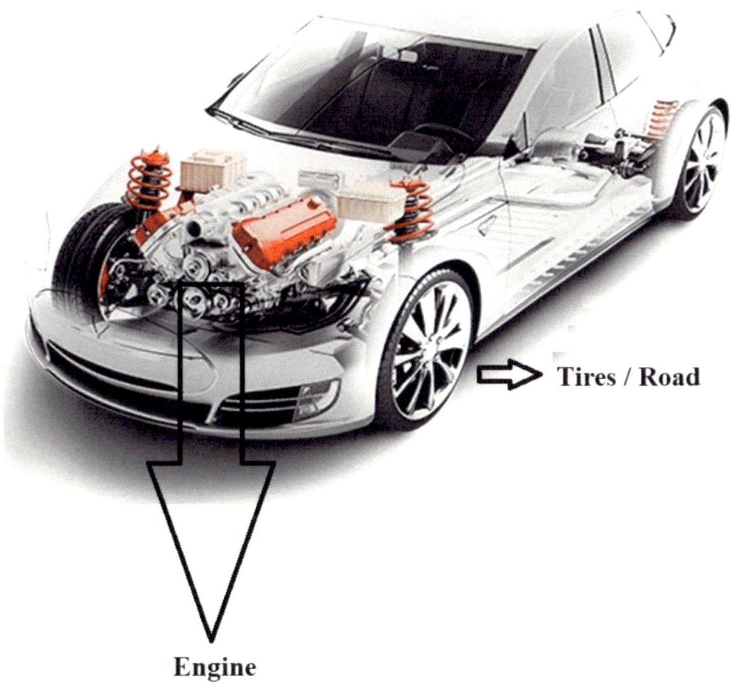

Figure 2.
Noise sources in a vehicle.

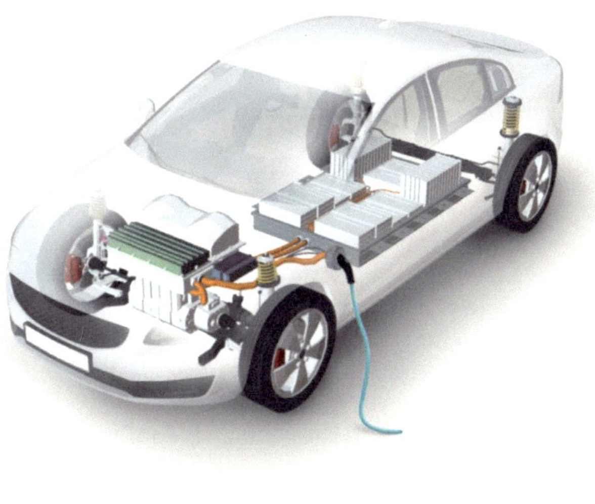

Figure 3.
Electric vehicle.

engine (ICE) vehicle can be greater than 6 dB (A) at 10 km/h. Unfortunately, much later at higher speeds, both types of car become equally loud, mainly due to tire noise.

When considering how traffic flow affects the subjective perception of noise levels, it can be concluded that it depends on the number of vehicles, their speed, and structure as described in the following paragraph.

A traffic flow of 2000 vehicles per hour produces twice the perceived noise level than 200 vehicles per hour. If the traffic speed is 105 km/h, it produces twice the perceived noise level than the 50 km/h traffic flow. One heavy weight vehicle (HV > 3.5 tons) with a speed of 70 km/h creates a perceived noise level of 28 light-weight vehicles (LV <3.5 tons).

2.2 Railway traffic noise

The main sources of railway traffic noise are noise generated from:

a. Vehicles traveling the railway.

b. Maneuvers.

c. Wagons.

d. Electromotor trains.

e. Motor trains.

f. Warning signals.

There are several other significant sources of noise, apart from the main sources mentioned above, which are:

- Propulsion systems for railway and railway vehicles.
- Interaction of wagon wheels, locomotives, and trains with rails.
- Braking process.
- Additional equipment such as ventilation, sirens, air-conditioning, and heating.
- Aerodynamic noise, especially in the case of high-speed trains.

On the propulsion system, noise is mainly generated by the operation of the traction engine (suction and exhaust process in the case of the diesel engine which is also the noisiest type of engine), the engine cooling system, the transmission system, and the ventilation system.

Wheel-rail interaction generates dominant noise in railway vehicles and depends directly on the speed of movement and the geometric configuration of the railway track. When moving on straight railway sections, the noise is mainly generated as a result of the roughness of the wheel and rail surfaces, i.e., from their friction. When driving through the railway curves, the wheels make more noise, not only due to rolling but also due to slipping of the metal wheels, which can be observed as creaking along the railway track. The cause of this phenomenon is the constructive nature of the wagons themselves, in which the wheels are fixed with parallel axles, which is why the outer wheels, when crossing a longer path than the inner ones, must glide, thus generating noise.

The noise generated by the braking process, in addition to the roughness of the wheels and track contact surfaces, depends significantly on the type and form of used brakes.

Noise from additional equipment is mostly generated by fans and their engines. Furthermore, it is important to mention the noise generated by the warning and notification signals.

Aerodynamic noise is caused by the passage of the train through the airspace. The noise level generated by air turbulence at or near the train surface in motion is logarithmically proportional to the train speed; therefore aerodynamic noise is significant only at higher speeds [3].

Figure 4 shows the noise sources of a high-speed train, apart from ground vibrations resulting from its passage and the conversion of structural sound to airborne sound in buildings.

2.3 Aircraft traffic noise

Aircraft traffic also causes several environmental problems or in other words an increase of noise. Nowadays, when observing the rapid development of all types of traffic, especially aircraft traffic, it can be concluded that there has been a significant increase in noise levels. In particular, the population living near airports is affected by the negative effects of noise exposure.

Aircraft noise can be divided into groups, which are shown in **Figure 5**:

- Noise caused by different types of engines.
- Noise caused by aircraft structure.

The noise sources generated by the engine groups are:

- Turbojet engine.
- Turbofan engine.
- Propeller propulsion (classic or turbine engine).

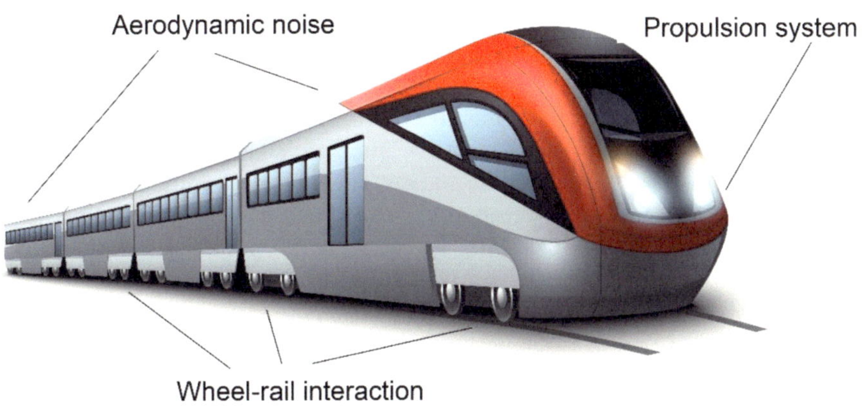

Figure 4.
Significant noise sources in the case of the high-speed train.

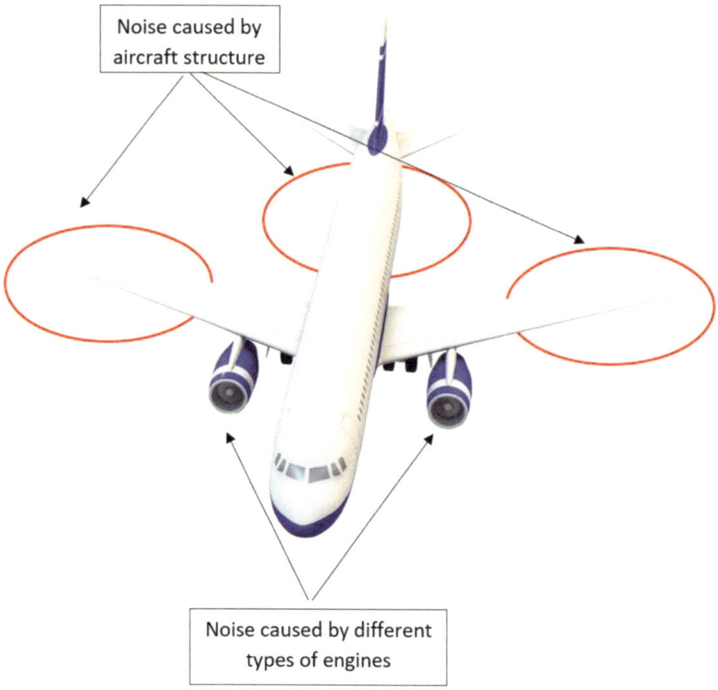

Figure 5.
Significant aircraft noise sources.

The noise produced by a turbojet engine can be divided into following groups:

- Compressor noise.
- Vibration-induced noise.
- Output jet noise.

Turbojet engine noise presented a big problem in the 1960s, especially the intake noise of this type of engine. The noise source of such intake noise is the compressor blades. During time as technology has evolved, aircrafts have become quieter, and therefore noise reduction in that sense continues even today.

The turbofan engine was designed to reduce aviation noise levels. In the case of the first turbofan engines, the largest noise source was the compressor, turbine, and jet exhaust. Newer turbofan engine models have succeeded to reduce the aforementioned noise levels. Turbofan engine consists of blades and a turbojet engine. This type of engine is often used in commercial aircraft industry.

Aircraft structure noise is defined as the sound which is produced from the movement of air between a solid body and air. The largest "manufacturers" of aircraft structure noise are landing gear, aircraft wings, and the flaps which are shown in **Figures 6** and 7. The noise generated by these aircraft parts depends on different aircraft configurations.

The noise level of an aircraft takeoff can be compared to the noise level produced by the engine group, while the landing aircraft engine group noise level is almost insignificant.

The noise produced by the flaps is created by the outer edges of the flaps. The main cause of flap noise is the emersion of air vortex which is created by flap extension. This vortex is the main cause of noise at the end of the wings.

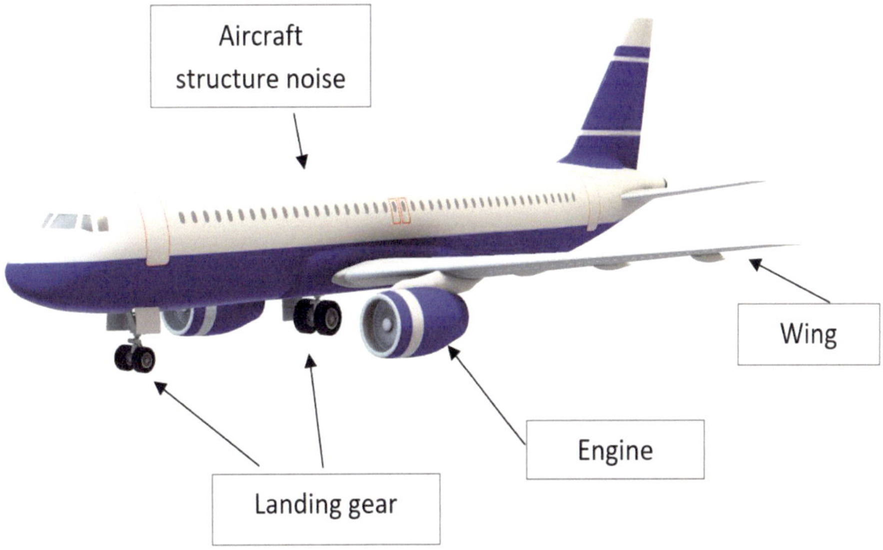

Figure 6.
Aircraft structure noise sources.

Figure 7.
Wing with extended flaps.

Another significant source of aircraft structural noise is landing gear. Landing gear noise is generated during takeoff and landing of an aircraft. During takeoff and landing, the landing gear is lowered; thus high air resistance occurs, which produces the landing gear noise [4].

2.4 Other types of noise sources

Other noise sources include industrial noise, noise caused by various construction work, and noise produced by different music and sports events.

Industrial noise (shown in **Figure 8**) is the amount of acoustic energy received by the human hearing system while working at the industrial hall. Occupational noise or industrial noise is a common term used when it comes to occupational

Figure 8.
An example of industrial site which produces industrial noise.

safety, since prolonged exposure to this type of noise can cause various health problems (e.g., annoyance, loss of concentration, sleep disorders, headaches, etc.). The worst consequence of prolonged exposure to this type of noise is permanent hearing impairment. Bearing in mind all the above, it can be concluded that this kind of noise certainly affects work efficiency.

When considering noise caused by different construction sites which are shown in **Figure 9**, this type of noise can have extremely high noise levels. Furthermore, such noise levels are very variable given that the construction process has many different phases. Thus, depending on the type and phase of construction, this category of noise can have indoor and outdoor noise sources and sometimes both at the same time. Activities on construction sites include the use of hammers, off-road trucks, cement mixers, cement cutters, electric saws, welding machines, as well as noise generated by hand tools such as a drill. Therefore, such noise represents a challenge for the workers and in addition for the population located near the construction site. This type of noise may have health consequences identical to those described in the previous section for the case of industrial noise.

Musical events are very dynamic (see **Figure 10**). In this case, the sound engineer plays a key role in ensuring that the audience gets the full experience of a music event by mixing the music. The order of the songs is usually strategically set in a way that higher levels of tempo or dynamics and energy remain until the end of the night, which represents a certain kind of "peak" of the concert. Naturally, the sound engineer will want to raise the sound levels as much as possible, so it can be

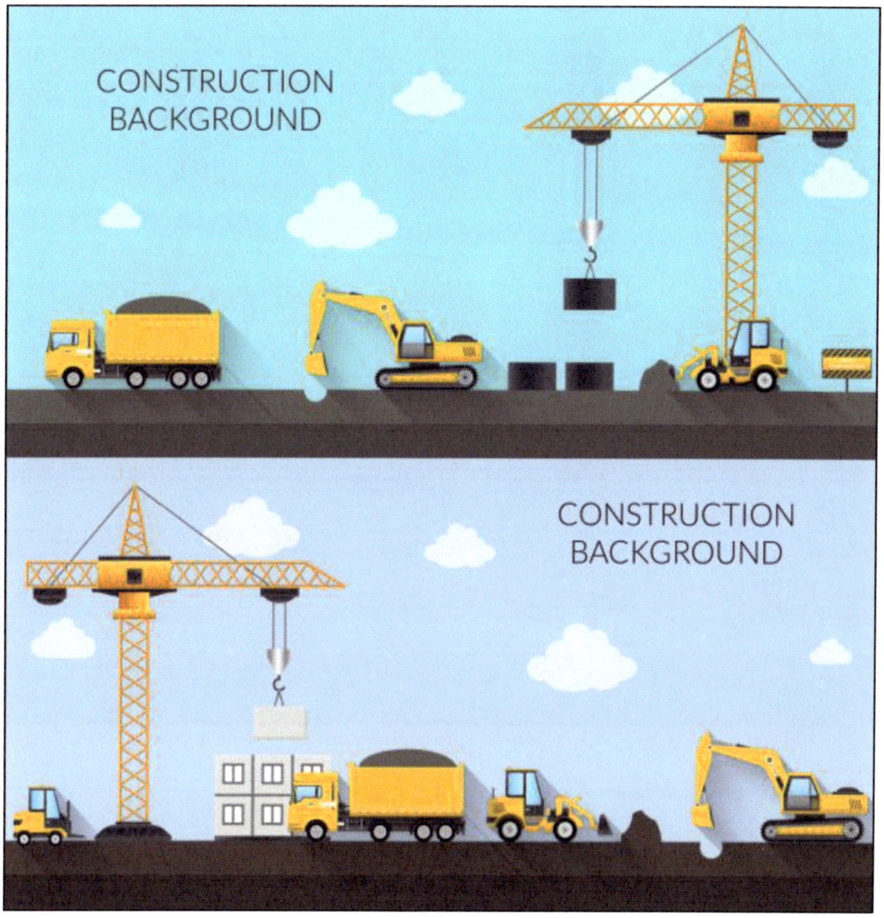

Figure 9.
An example of construction site and some typical noise sources.

expected that the noise levels will increase as the night passes. In addition, stage orientation plays a significant role in sound propagation. If the concert takes place outdoors, the reality is of course that people who are not actually present at the concert site, however live near, will hear the music. In that case it has to be noted that the music impact will be minimal at a distance of more than a mile or two from the concert site, so this type of noise could be annoying or unpleasant (especially if one does not prefer the music performed by an artist). The concert will certainly not be suspended due to a complaint from only one person living relatively near the concert site. Licensing of open-air concerts by the competent authority is a well-established process. Therefore, one can expect only a few concerts a year from a particular outdoor site. Concert organizers can in addition send notices to home-owners near the concert site reminding them of concert details, curfew time, and their right to complain if noise levels become significant and therefore annoying.

Sport events (shown in **Figure 11**) present a very similar situation as music events. Although most people enjoy them, those who are disturbed by the noise levels produced can be protected in some way by using different types of ear protection (e.g., noise-cancelling headphones or popularly called earbuds). For people who are particularly sensitive to noise, there still remains the option of simply physically moving away for a while from the site where a particular sporting or musical event will take place.

Figure 10.
An example of open-air musical event.

Figure 11.
An example of sport event.

3. Ten ways how to reduce noise levels

The previous sections of this chapter have described the most common traffic sources in urban areas. The aim of this paragraph is to propose and describe measures to reduce such noise which are shown in **Figure 12** [5].

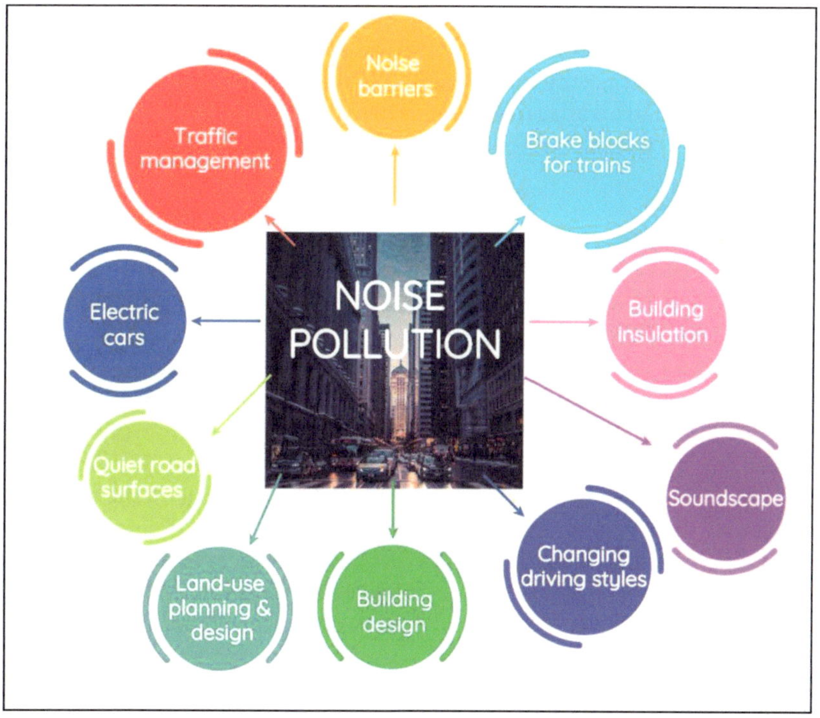

Figure 12.
Ten ways to reduce noise pollution based on [5].

Ten ways to reduce noise in urban areas proposed in [5] are:

- Urban planning.
- Designing living spaces.
- Sound insulation of living spaces.
- Smart traffic management.
- Implementation of quiet road surfaces.
- Development of train brake blocks.
- Electric cars.
- Changing driving styles.
- Noise barriers.
- Application of soundscape concept.

It is important to emphasize that these solutions are not the only solutions and that there are still different opportunities and prospects for progress and development of both existing and new methods.

In the following sections, a more detailed explanation on how electric vehicles affect the reduction of noise levels will be provided, especially in urban areas. On the other hand, problems which occur with electric cars will be discussed. In addition, the effect of smart traffic management system, traffic behavior changes, and quiet road surfaces in terms of noise reduction will be examined.

3.1 Electric vehicles

Electric vehicles (shown in **Figure 13**) present the future in terms of reducing noise pollution in urban areas. Electric vehicles are quieter especially when compared to vehicles with an internal combustion engine.

Electric vehicles at low speeds produce very low levels of noise, i.e., in current urban environments, these vehicles are practically silent and unnoticeable. For example, the difference in noise level between an electric vehicle and an internal combustion engine vehicle can be greater than 6 dBA at 10 km/h [6]. At higher speeds, both types of vehicles become equally loud, mainly due to the tire noise. In urban areas, for pedestrians (especially for vulnerable groups: children and visually impaired people), it becomes much more difficult to detect electric vehicles due to their aforementioned lower noise levels [6]. Therefore, it is necessary to find a solution in the form of an audible signal that electric vehicles will emit in different driving modes.

Since 2009, the Japanese government, the United States Congress, and the European Commission have been studying the legislation to determine the minimum level of emitted sound signal for plug-in electric and hybrid vehicles when operating in electric mode. This level of audible signal must be such that visually impaired people, other pedestrians, and cyclists can hear the electric vehicles in motion and detect from which direction they are coming from. Several tests and studies have shown that vehicles operating in electric mode below 32 km/h are almost inaudible for pedestrians [7].

In 2011, the European Commission composed guidelines for Acoustic Vehicle Alerting Systems (AVAS). The aim of the guidelines was to present

Figure 13.
Electric vehicles: the future.

recommendations to manufacturers for a system to be installed in this type of vehicles that would emit an audible signal to pedestrians and other vulnerable groups in traffic. The guidelines recommend that AVAS automatically generate a continuous sound in the minimum range of vehicle speeds from standing at a place (0 km/h) and starting to drive (up to approximately 20 km/h) and when driving backwards, if applicable to that category of vehicle. Furthermore, the guidelines suggest which types of sounds are not suitable for this purpose [8]. In February 2013, the European Parliament decided that the law draft should combine series of tests, norms, and measures that first must be developed in order to make AVAS mandatory in the future. The approved amendment stipulates that "the sound generated by the AVAS should be a continuous sound of the vehicle in operation providing information to pedestrians and vulnerable traffic users. The sound should clearly demonstrate the behavior of the vehicle and should sound similar to the sound of a vehicle of the same category equipped with an internal combustion engine" [9]. In April 2014, a law (Regulation (EU) No 540/2014) was approved by the European Parliament requiring AVAS to be mandatory for all new electric and hybrid vehicles. The new guidance proposes a transitional period of 5 years after the announcement of the final approval of the April 2014 proposal [10].

For example, a case study was carried out in Zagreb in 2019 [11], which involved 201 participants who had the task to fulfill a specially designed questionnaire. This case study addresses the issue of electric cars in everyday traffic. The research was focused on assigning a desirable (both for pedestrians and drivers) and, at the same time, detectable warning sound to an electrical vehicle in the daily traffic. The case study showed that the majority of participants (especially the ones with a driving license) would prefer that their electric vehicle sounds like an internal combustion engine car. The "nondrivers" were more open to the solution that an electric vehicle has a different sound than a "regular" car. According to the study, they were more opened to a solution of adding a sound of an electric motor to the electric vehicle as a warning sound which would distinguish the electric cars from cars with internal combustion engine in everyday traffic. However, an important question concerning the overall quality of life remains: "Which one of these two sounds would increase more the noise levels in urban environments?"

Finally, it can be concluded that electric vehicles will play a significant role in reducing noise levels especially in urban areas while adequately addressing the problem of emitting a certain warning sound when parking, moving forward, and stopping. It is important to note that the unique warning sound has not yet been implemented, i.e., various car manufacturers are still "experimenting" regarding this issue.

3.2 Smart traffic management

Smart traffic management is a system in which centrally controlled traffic signals and sensors regulate the flow of traffic through the city in compliance with the current state on the roads in the city (see **Figure 14**).

Upgrading and integrating all the signals on major roads in the city will have multiple benefits such as:

- Significant reduction of daily traffic congestion, equalization of traffic flows, and prioritization of traffic in response to real-time demand.
- Pollution reduction in the city: stop-start driving is inefficient and polluting.
- Providing priority for busses approaching intersections and phase-coordinating traffic lights enabling a "green wave" through the city.

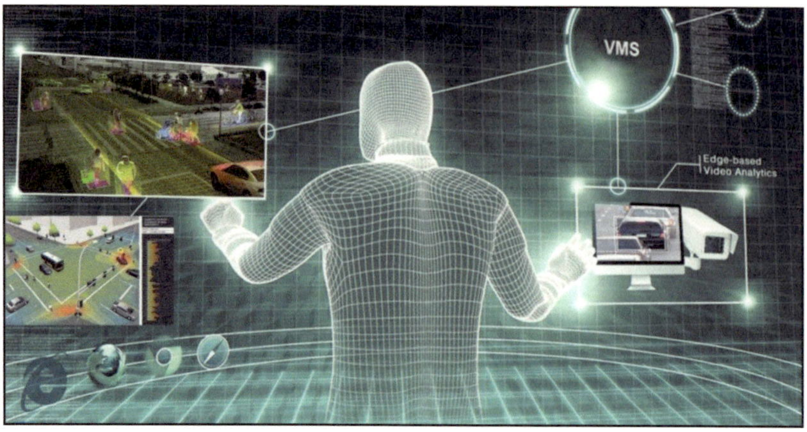

Figure 14.
Smart traffic management system.

- Enabling a much more efficient response to traffic accidents, especially on motorways, for example, the system can be pre-programmed for a sudden increase in traffic.

- Enabling inbound traffic flow control.

In addition to the multiple benefits listed above, the system would also provide the perfect opportunity to install tracking equipment and collect a much more detailed traffic and travel data. Each set of traffic lights would have communication equipment that can be used to transmit (anonymously) vehicle data, either from automatic number-plate recognition (ANPR) cameras or Bluetooth detectors and closed-circuit television (CCTV) transmission (if suitable). There are three components in smart traffic management: traffic lights, queue detectors (in terms of traffic congestion) embedded in the road, and cameras and a central control system. Queue detectors define the traffic flow control system on all major roads in the city. The system controls the traffic lights to maintain the free flow of traffic within the city. Every 2 seconds, the system uses a real condition model to decide whether one will have the priority of changing the phase of any of the traffic lights. A system software considered as an "asset" can be defined as, for example, obeying the bus timetable, less pollution at a particular location, or fewer vehicles waiting at a highway toll booth.

If inbound traffic flow control is used, the most remote sets of traffic lights on arterial or radial roads serve as a special function and are technically known as "doors" or "control points." They regulate the flow of vehicles entering the city.

One example of software with the purpose of smart traffic management is split cycle and offset optimization technique (SCOOT) which is used in hundreds of European cities for decades. It is used in Cambridge for coordinating traffic signals, where it usually favors busses. In Zurich, Braunschweig, and Potsdam, the system is used to control all traffic in the city [12]. The software is deployed with "knowledge" of the road network and is trained to respond appropriately to a wide range of scenarios (e.g., major traffic "disruptions" such as an accident on the arterial roads). It is important to note that the system also has the option to manually manage and make changes if there is a need for it.

3.3 Changing driving styles

Traffic behavior psychology is defined as the study of the behavior and psychological processes of different traffic participants. Its aim is to attempt to identify specific behavior patterns of users of different types of traffic with the ultimate goal of developing effective anti-accident measures [13]. There are two basic approaches that can help psychologists develop and implement measures against traffic accidents. First, traffic psychology can act as an "assistant" of science with a dominant field of traffic engineering. Road safety engineering solutions aim to optimize internal road safety. A safe road can be defined as a road that is designed, operated, or modified in such a way that it [14]:

- Warns the driver of any unusual or odd features.
- Informs the driver of road conditions.
- Guides the driver through atypical parts.
- Controls the passage of drivers through problematic points and roads ("black" traffic points).
- Has the ability to tolerate a driver's impolite or inappropriate behavior.

Engineering is powerful for a significant number of traffic problems. However, it would be wrong to assume that it is exclusively an engineering solution. Engineering must also consider sociopsychological solutions that include the implementation, education, and other activities in order to change the behavior of road users. In a significant number of traffic situations, psychological measures can support engineering measures in such a way that the performance of expected safety works even more effectively by informing or motivating traffic participants to change their behavior in the desired direction.

Regarding the specific application of this topic to the issue of noise, changes in the traffic participants' behavior would mean a complete "openness" to newly developed traffic monitoring systems, participation in them, and raising awareness of the most vulnerable groups in traffic (visually impaired people and children). **Figure 15** shows worrying data which is a direct consequence of the current behavior of road users.

3.4 Quiet road surfaces

In previous paragraphs, it has already been established that the dominant noise source when driving a car at higher speeds is tire noise which is caused by friction between the wheels and the road surface. In the case of light vehicles, tire noise becomes the main noise source already at a speed of 30 km/h, while in the case of heavy vehicles at speeds higher than 60 km/h tire noise becomes the main noise source, which is shown in **Figure 16** [16]. **Figure 17** shows noise levels for different types of vehicles depending on their speed [17].

Tire noise depends on the following road surface properties:

- Surface texture.
- Acoustic absorption.
- Aerodynamic processes.

Improving road surface properties in a way that effectively reduces noise generation and amplification will result in lower noise levels. There are several types of quiet road surfaces, and their application is mainly determined by the noise

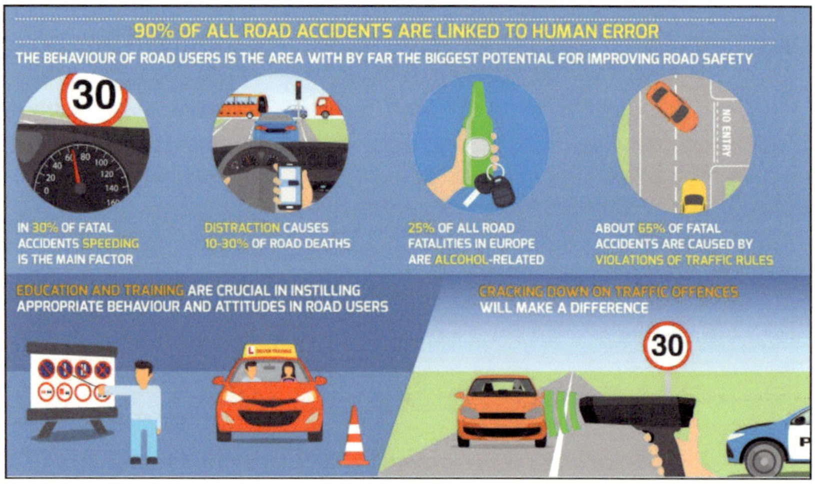

Figure 15.
Review of irresponsible and inappropriate traffic participants' behavior [15].

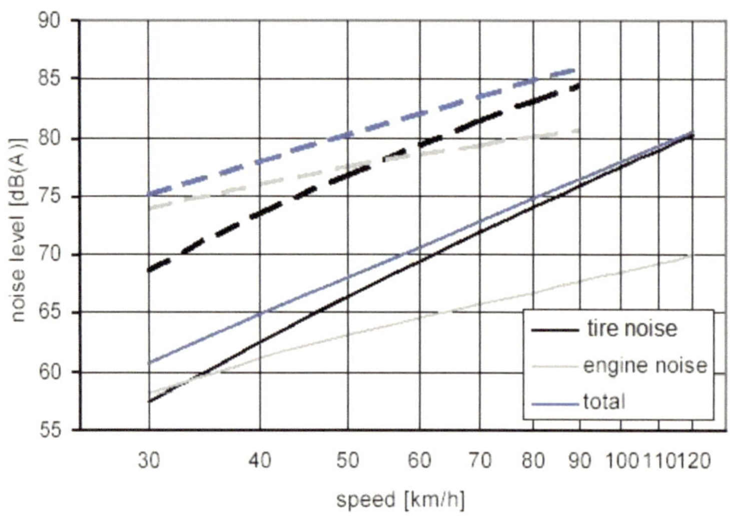

Figure 16.
The correlation of noise levels and vehicle's speed (lightweight vehicles marked with full lines and heavy weight vehicles with dashed lines) [16].

reduction proportion, the permitted speed in traffic, the composition of the traffic flow, and the possible adhesion of tires to the surface during parking. In urban areas, three types of bases are most commonly used:

- Thin surface layers.
- Two-layer porous asphalt.
- Cast asphalt.

Thin surface layers are often referred as thin asphalt layers of or thin asphalt bases for noise level reduction (see **Figure 18**) [1]. These layers are usually up to 3 centimeters thick. There are a significant number of different types of thin surface

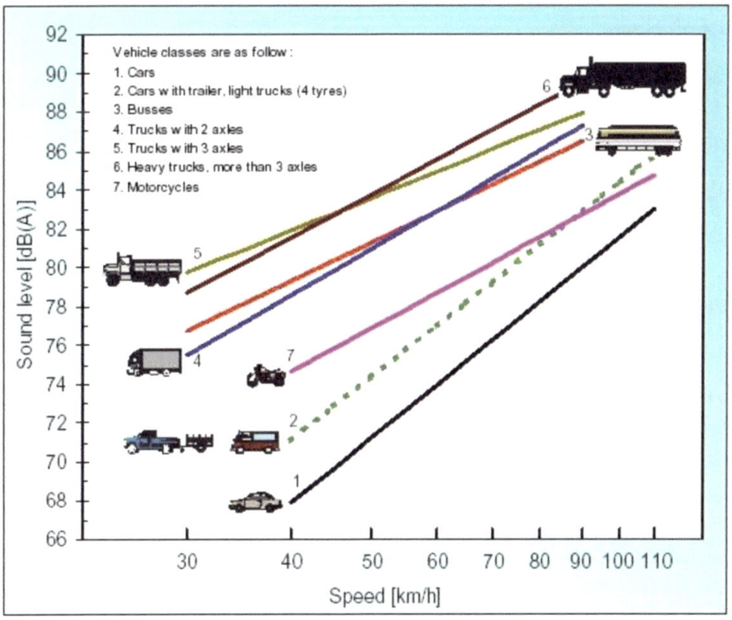

Figure 17.
Noise levels for different types of vehicles, depending on their speed [17].

Figure 18.
Two-layer porous asphalt (on the left) and a thin surface layer of asphalt (on the right) [1].

layers on the market, for example, in the Netherlands more than 40, including porous and dense types. They usually reduce noise by 2–4 decibels at 50 km/h for cars when compared to the average dense asphalt concrete. Porous asphalt types are in average about 1 decibel quieter than dense ones; however they have a shorter duration than dense asphalt. The typical duration of a thin surface layer is 7–9 years. Thin surface layers are suitable and increasingly popular on low- and medium-speed roads; however they are not appropriate for places exposed to strong stress forces, such as roundabouts, steep slopes, bends, truck exits, etc.

The two-layer porous asphalt consists of a top layer (2.5 centimeters thick) and a lower layer (4.5 centimeters thick) which is shown in **Figure 18**. The total thickness of the 7 centimeter porous layer absorbs more noise or more precisely at the beginning of its implementation from 5 to 7 decibels. Two-layer porous asphalt is relatively expensive and suitable for high-speed roads that require extreme noise reduction. Cast asphalt has a thin (3 centimeters) surface layer with a specific molding design. It contains more stone than thin surface layers, and since it is not porous, it does not absorb as much noise; however it is more robust than other

Tire size [inches]	Sound pressure level [dB(A)]
<145	72
145–165	73
165–185	74
185–215	75
>215	76

Table 2.
Table of decibels [18].

asphalts. A test of this type of cast asphalt conducted in Berlin resulted in an initial noise reduction of 1.5 decibels. In addition to installing quiet road surfaces, another method of reducing tire noise is the production of quiet car tires. There are several manufacturers that have developed such tires and successfully placed them on the market. In general, the comfort concept of tires is directly related to their loudness. One of the tire functions is to absorb impacts and dampen vibrations, which means that the tire is an element of the vehicle that ensures the travel comfort. Smaller wheels produce less noise. Basically, a smaller tire represents a smaller surface that adheres to the road and thus produces less noise. In addition to the size, the material from which the tire is made is also significant. There are softer types of rubber that also make less noise. Of course, one of the most important factors is the speed of driving. If one plans to drive at higher speeds, it makes sense to have tires with such performance. However, such tires are thicker and larger, thus creating more noise. Furthermore, weather conditions also play a key role in choosing tires. Tires selected for severe weather will create more noise due to certain safety aspects, i.e., the need to better adhere to the road surface. Tires selected for extreme weather conditions will make the highest noise level. According to their design, tires selected for city driving can make less noise. All of this logically implies that winter tires will make more noise than do summer tires.

Tire manufacturers can produce tires that make less noise. There are already various models, especially the quietest summer tires, which are 4–6 dB(A) below the limit, and many of the newer winter models are also approaching the limit of 2 dB(A). These restrictions are determined by Regulation No. 117 United Nations Economic Commission for Europe (UNECE)—Uniform requirements concerning the approval of tires regarding the emission of rolling sound and/or traction on rainy surfaces. The sound pressure levels generated by each tire size are shown in **Table 2** (for reinforced tires (XL), the limits are higher by 1 dB(A)) [18].

It can be concluded that noise can be limited by using modern quiet road surfaces which reduce its level from 3 to a maximum of 7 decibels. Unfortunately, such materials are usually 2.5 times more expensive than ordinary materials. Furthermore, noise can be reduced by using quieter tires by an additional 3–4 decibels; however the choice of tires depends on the preferences and habits of the driver. In the majority of cases, noise is reduced by speed limits on local roads and highways located near resident areas, and these restrictions are often even more restraining at night.

4. Conclusions

Noise pollution is a serious problem that affects the overall quality of life. This problem is especially noticeable in urban areas where a significant amount of noise pollution is produced by traffic. In this chapter the main traffic sources are described and analyzed. In addition to road, railway, and aircraft noise sources, other typical noise sources common for urban areas are also discussed. Bearing in mind the serious consequences of long-term exposure to noise, it is necessary to implement at least some measure to reduce noise levels. Today there are many initiatives and plans how to tackle this issue; however this chapter has focused on measures directly connected to traffic noise levels. In that sense, this type of noise reduction measures has been described and discussed in detail.

Furthermore, it can be concluded that education and some form of encouragement are needed to get the people more involved in the "fight" against noise and its negative impact. In this way, a kind of pressure would be created to set up the necessary city infrastructure (sensor networks), and finally the citizens would obtain a much-needed improvement of the quality of life in the environments in which they live.

Author details

Sanja Grubesa* and Mia Suhanek
Faculty of Electrical Engineering and Computing, Department of Electroacoustics, University of Zagreb, Zagreb, Croatia

*Address all correspondence to: sanja.grubesa@fer.hr

References

[1] Helene A, Klæboe R. A Nordic Perspective on Noise Reduction at the Source. TØI report806; 2005

[2] Braun ME, Walsh S, Horner JL, Chuter R. Noise source characteristics in the ISO 362 vehicle pass-by noise test: Literature review. Applied Acoustics. 2013;**74**(11):1241-1265. DOI: 10.1016/j.apacoust.2013.04.005

[3] Thompson D. Railway Noise and Vibration Mechanisms, Modelling and Means of Control. Amsterdam, The Netherlands: Elsevier Science; 2008

[4] Bertsch L, Simons DG, Snellen M. Aircraft Noise: The Major Sources, Modelling Capabilities, and Reduction Possibilities. Report Number: DLR IB 224-2015 a 110. German Aerospace Center & Delft Technical University; 2015

[5] SfEP Future Brief. Noise abatement approaches [Internet]. 2017. Available from: https://ec.europa.eu/environment/integration/research/newsalert/multimedia/infographics/10_ways_to_combat_noise_pollution_standalone_infographic.pdf [Accessed: 05 May 2020]

[6] Parizet E, Pondrom P, Janssens K. Additional efficient warning sounds for electric and hybrid vehicles. Applied Acoustics. 2014;**86**:50-58

[7] King M. Hybrid cars not noisy enough, group says. In: The Gazette (Montreal); 2010

[8] Working Party on Noise (GRB) Quiet Road Transport Vehicle Working Group. Annex VIII - Proposal for guidelines on measures ensuring the audibility of hybrid and electric vehicles. United Nations Economic Commission for Europe; 2011

[9] European Parliament. European Parliament legislative resolution of 6 February 2013 on the proposal for a regulation of the European Parliament and of the Council on the sound level of motor vehicles (COM(2011)0856 – C7-0487/2011-2011/0409(COD)). European Parliament. Amendments 16, 20, 58 and 59; 2013

[10] European Commission Press Release. Commission Welcomes Parliament Vote on Decreasing Vehicle Noise. European Commission; 2014

[11] Suhanek M, Djurek I, Petosic A. A Case Study, The Urban Residents' Choice for Electric Vehicles Warning Sounds. Special Issue "Smart Cities – Innovative Approaches" within the American Journal of Environmental Science and Engineering; 2019; pp. 47-51. DOI: 10.11648/j.ajese.20190303.11

[12] Strehler M. Signalized flows optimizing traffic signals and guideposts and related network flow problems [thesis]. Senftenberg: Brandenburg University of Technology Cottbus; 2015

[13] Rothengatter T. Psychological aspects of road user behaviour. Applied Psychology. 2008;**46**(3):223-234. DOI: 10.1111/j.1464-0597.1997.tb01227.x

[14] Goldenbeld C, Levelt PBM, Heidstra J. Psychological perspectives on changing driver attitude and behaviour. In: Recherche - Transports – Sécurité. 2000. pp. 65-81. DOI: 10.1016/S0761-8980(00)90108-0

[15] Road safety facts [Internet]. Available from: https://roadsafetyfacts.eu/how-can-road-users-improve-safety/ [Accessed: 05 May 2020]

[16] Dijkink JH, van Keulen W. Silent Transport – An Innovative Silent Road Concept for Heavy Vehicles. Internoise

2004, Prague, Czech Republic; 2004 Paper No. 767

[17] Sandberg U. Work Package 1.1. Vehicle Categories for Description of Noise Sources. HARMONOISE. Contract Number: IST 2000: 28419. 2003

[18] Tyre loudness [Internet]. 2019. Available from: https://www.oponeo.co.uk/blog/tyre-noise [Accessed: 05 May 2020]

Section 2

Numerical and Experimental Application for Noise Reduction

Chapter 6

Noise Transmission Losses in Integrated Acoustic and Thermo-Fluid Insulation Panels

Himanshu Dehra

Abstract

A simulation model is proposed for integrated acoustic and thermo-fluid insulation constituting an airflow window with a photovoltaic (PV) solar wall spandrel section. The physical model of an outdoor test-room comprises of a wooden framed double or cavity wall assembly with: (i) a triple glazed fenestration section with a closed roller blind; (ii) a solar wall spandrel section of double-glass PV modules and back panel of polystyrene filled plywood board; and (iii) fan pressure-based manually operated inlet and exhaust dampers with ventilation through an exhaust fan for transportation of heat. A generalized two-dimensional analysis of a double wall structure is illustrated by the placement of surface and air nodes into two adjacent stacks of control volumes representing outer and inner walls. The integrated noise insulation and energy conversion model is presented. The energy conversion and noise insulation model are supported with some numerical results using devised noise measurement equations. The following additional parameters are also calculated to support the integrated insulation model: noise transmission losses and noise reduction coefficients for various types of noises. State-of-the-art of acoustic and thermo-fluid insulation along with general building construction guidelines for acoustic and thermal insulation are also presented.

Keywords: integrated noise insulation, insulation materials, acoustic insulation, thermo-fluid insulation, solar energy

1. Introduction

The passage of air in and out—and heat along with it—is called infiltration. The significance of infiltration is clear from the fact that the average house has 600 cm^2 of vents and flues alone; plus window frames, doors, sills, and corners that need sealing and plugging; fireplaces with unfit dampers; and a front door that is slammed 3000 times a year. Air passes in and out through these openings. Therefore, infiltration is a misleading word—since it only denotes a one-way movement of air into a house. There is need of better word than breathing or respiration or exchange. Focusing entirely on insulation thickness, the best way to define all flow of air in and out through unsealed slits and unplugged holes is by huff and puff of a house. Due to a slit between the sash and the frame or the frame and the house, a 1.25 × 1.25 m^2 double glazed window will easily lose about 70 L of oil in winter, and some studies have suggested that windows account for even more heat

IntechOpen

loss than roofs. Double or triple panes are not the only solution. Traditionally, the most effective defense is solid, thermal shutters, put up from the inside that fit the window hole exactly.

This chapter attempts to bring up integrated noise insulation modeling for airflow windows along with their use as a sustainable energy source of electric and thermal energy besides providing daylighting. The airflow window system with photovoltaic modules embedded in solar wall spandrel as illustrated in **Figure 1** has been considered for investigations [1–60]. A triple glazed airflow window combined with a PV solar wall spandrel has many advantages: (i) airflow window provides electric power, hybrid ventilation through heating/cooling, daylighting and reduction in greenhouse gas emissions by energy conservation of fossil fuels; (ii) it provides integrated sound and thermal insulation; (iii) it gives protection from excessive heating from solar radiation by passing and controlling the amount of heat transport; (iv) it gives protection to PV modules from excessive heating, weather deterioration including protection from snow and dust; (v) with frame supporting structures, the system is easily approachable for repair and maintenance jobs; and (vi) it has better esthetic appearance to the occupants and to the viewers from outside in comparison with stand-alone PV module power generating system.

1.1 Physical model description

A full-scale experimental test section comprising an airflow window with a single pane exterior glazing and a double pane interior glazing and a photovoltaic solar wall spandrel was constructed in an outdoor room facility at Concordia University, Montréal, Québec [3–60]. The transportation of heat was achieved through manually operated intake and set of exhaust dampers. The intake and exhaust dampers located on exterior side were fitted with wire mesh screens. The exhaust damper placed on the interior side toward outdoor room is for allowing the pre-conditioned outdoor air directly into the room indoor environment. The air movement in the airflow window system was achieved through: (i) buoyancy-induced hybrid flow and (ii) fan-assisted flow created either in absence or presence of wind-induced flow.

In airflow window integrated with PV module, there are many issues related to its operation. The main objective is to model the integrated thermal and sound

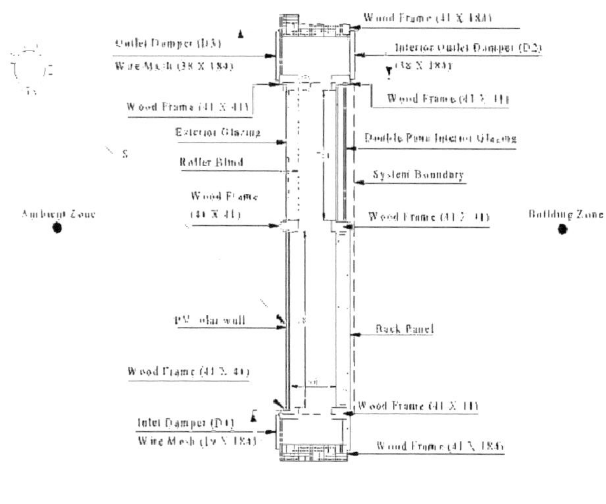

Figure 1.
Schematic of an airflow window system with a PV solar wall spandrel section (dimensions provided are in mm).

insulation fields and to maximize the value of total energy generated and therefore increase the combined efficiency of the system. In achieving this objective, the heat transfer model and losses of PV model need to be studied. The heat can be recovered from the heated PV modules in different ways. One of the options is to treat the heated PV module as absorber surface and pass airflow through the surface. The key operation parameters are air mass flow rate, area of the absorber surface (PV, Blind) and its temperature. Air mass flow rate requires optimum air velocity under various conditions to prevent condensation, stagnant zones, flow reversal and other adverse effects that will cause deterioration of the PV panels. This optimization of airflow will maximize the absorbed solar radiation conversion to either thermal or electric power. By placing motorized blind after the integrated PV, will achieve twin objectives, control the amount of daylight transmission and also help in absorbing the excess external heat, which is going inside the room and overheating it. This heat is further taken from blind by mass flow of air, while cooling the PV modules. Depending on the climatic variables the operation of airflow window can be optimized. Damper motion is controlled by both flow and temperature sensors. During the summer, in the daytime, when outside air temperature is in the range of 15–22°C, the exhaust air vent provided will exhaust the heated air from PV modules to outside. But during night, the cooled fresh air can be used inside the room to reduce the cooling load of the building. Another important aspect is to control the amount of daylight transmission by optimizing sizing and distribution of opaque photovoltaic modules for maximizing the value of thermal and electric power generated.

2. State-of-the-art

Traditionally, insulation is dead air space, or a dead gas space, sometimes combined with a reflective surface. Air has a low inherent conductivity. If it is dead or motionless there is no convection and when there is a reflective surface, radiation is cut to a minimum. Dead air space can be found in materials like fiberglass blanket, loose-fill and foam. They embody thousands of tiny pockets where dead air is encased, protected and preserved.

Outdoor duct system consists of combination of fans, duct construction elements, heat exchangers and air filters. The location of fan in outdoor duct is decided by the direction of flow and desired pressure relations. A supply fan is used to pump air into a space and exhaust fan is used to draw air out of a space. The pressure relations for the two cases are different. There is a buildup pressure in case of supply air and reduction of pressure in case of exhaust air. The space pressure is determined based on the relative quantity of air handled by supply and exhaust air. The space pressure will be positive if there is an excess supply air in comparison to exhaust air and negative if there is an excess exhaust air in comparison to supply air.

The total energy provided by the fans to the air passing through a given system is fixed, assuming the same capacities and end pressures for a supply fan, an exhaust fan or both are used. A motor-driven fan is used to circulate filtered heated air from outdoor duct through supply duct to outdoor test-room. As the hot air is delivered through the ducts and into the room through the supply outlet, cooler air from the room is being returned through return grilles, into the exhaust air duct.

The choice of fan is dependent on creating a sufficient pressure is achieved to overcome the total losses based on the flow through the duct with longest run. Fan performance must meet and match system requirements. The only possible operating characteristic points are those where the system curve intersects the fan curve. These are the points where the pressure developed by the fan exactly matches the system resistance, and the flow in the duct system equals the fan capacity. The overall system resistance is calculated by summing the pressure losses for the individual components along any one flow path.

Air passing through the outdoor duct system is either heated or cooled. It results in decreased or increased density, respectively, and with assumption of constant mass flow rate, its total volume rate will be increased or decreased accordingly. The developed resistance in the duct is calculated based on actual gas density, volume, and velocity through it. The total system resistance is calculated by summation of these resistances along the flow path.

The difference in power requirements at various locations can be calculated. For the case of constant system resistance, the pressure required for the fan (P2 – P1) will be constant regardless of the location. The volume of air flow (Q) will vary with the location. The fan location with least power requirements is that place where density of air will be highest assuming constant efficiency.

Exterior duct insulation can be attached with adhesive, with supplemental pre-attached pins and clips, with wiring or bands. Liners can be attached with adhesive and supplemental pins/clips. Rectangular ducts and fittings are fabricated by grooving, folding, and taping with metal accessories such as turning vanes, splitters, and dampers incorporated into the system. If rectangular ducts exceed the pre-determined dimensions for particular static pressure, the ductwork must be reinforced. Insulation can significantly reduce operating costs that depend upon unit cost of heating and cooling energy, extent of duct exposed to outside conditions. In addition, duct insulation maintains the supply air temperature unaltered thereby, maintaining the conditioned space within acceptable temperature range. Vapor retarders are required on exterior insulation of ducts that are used for alternate heating and cooling.

Some thermal insulation materials can also serve purpose of sound control [1]. Acoustic efficiency depends upon physical structure of the material. Materials with open, porous surfaces have sound absorption capability [2]. Those with high density and resilient character can be used for absorption of vibrations. Insulation for sound conditioning includes flexible and semi-rigid, formed-in-place fibrous materials and rigid fibrous insulation. Thermal insulation materials improve their sound insulation when installed with discontinuous construction. A wall of staggered stud construction that uses resilient clips or channels on one side of the stud or resilient boards of special manufacture to prevent acoustic coupling mechanically between the surfaces, reduces sound transmission. Sound absorption by thermal insulation blanket in a cavity wall reduces sound transmission.

The energy conversion and noise characterization in an exterior double wall is important, for example, in modeling PV solar wall and transpired unglazed structures [1–60]. The energy conversion in a double of cavity wall is a function of solar irradiation, air gap width, mass flow rate and pressure, wall and air temperatures. A generalized two dimensional thermal analysis of an outdoor duct is presented by placement of surface and air nodes into two adjacent stacks of control volumes representing outer and inner walls of duct. A matrix solution procedure is adopted by constituting conjugate heat exchange of conduction, convection, radiation and ventilation heat transport.

The requisite amount of ventilation air in a building in a given climate depends on heating/cooling load. The HVAC load on building varies with the condition of outdoor ventilation air that may require additional heating, cooling and humidification or dehumidification. In temperate climates, outdoor air is more economical to use than recycled return air. The analysis of double or cavity wall for ventilation purposes using airflow window with PV solar wall structures is investigated. The investigation of energy conversion, ventilation and integrated insulation system is based on complete information for the given design conditions and limitations of operation results.

2.1 Absorption of sound

When a sound wave strikes a surface, part of its energy is absorbed by friction, part of its energy is transmitted, and the remaining part of its energy is reflected. But as reverberation directly depends on the loss of energy of sound wave due to friction, it is of greater importance. This property of a surface by which sound energy is converted into other form of energy is known as absorption and absorption coefficient of a surface indicates the degree to which this surface affects the absorption of sound. It is thus the ratio of energy absorbed by the area to the energy striking the area. The value of coefficient of absorption will depend on the frequency of sound. **Table 1** gives the value of coefficients of absorption for some common surfaces. These values correspond to the normal frequency of 500 cycles per second. It may be noted that coefficient of absorption for an open window is taken as unity. This is very easy to understand as sound wave approaching an open window must completely pass through it.

Sound absorbent materials: most of the common building materials absorb sound to a small extent and hence, for better acoustical requirement, some other materials are to be incorporated on the surfaces of the room. Such materials are known as absorbent materials and they help a great deal in making the room acoustically good. The important characteristics of absorbent materials are:

1. An ideal absorbent material should be economical in construction and maintenance, waterproof, fire-proof, sufficiently strong and good in appearance.

2. Noise level of the room provided with absorbent materials is considerably reduced.

3. In the hall treated with absorbent materials, speech can be heard clearly, and music can be fully enjoyed.

4. All the absorbent materials are found to be soft and porous. They work on the principle that sound waves penetrate into the pores, and in this process, sound waves are converted into other form of energy by friction.

5. The absorbing capacity of the absorbent materials depends on the thickness of the material, its density and frequency of sound.

6. The sound properties of the absorbent materials are considerably changed by their modes of fixing. Suspended absorbers in the form of inverted cones may be provided in the ceiling to make the hall acoustically sound good.

7. There is no royal road for making a particular room acoustically good. It mainly depends on the ideas of the technical staff either engineer or architect. Each case is to be studied separately and after proper thinking, suitable absorbent materials may be specified.

8. Great care should be exercised while prescribing the covering for an absorbent material so as to improve its appearance. Improper covering destroys the absorbent properties of the material.

Material	Absorption coefficient per m^2
Open window	1.00
Ventilators	0.10 to 0.50
Brick wall 40 cm thick	0.03
Plaster on wall surface	0.02
Glass against solid surface	0.03
Marble	0.01
Stage curtain	0.20
Linoleum or concrete floor	0.03
Solid wooden floor	0.09
Framed wooden floor	0.13
Plywood on battens	0.17 to 0.26
Window glazed	0.18
Curtains in heavy folds	0.40 to 0.75
Metal	0.01

Table 1.
Absorption coefficients.

9. It should be remembered that in a big hall, audience is a major absorbing factor. This is especially true in the high frequency zone. Hence, low frequency absorbent materials should be provided to achieve optimum reverberation time over a wide range of frequency of sound.

2.1.1 Types of absorbent materials

Various types of absorbent materials are available in the market under different trade names. The value of coefficient of absorption is supplied by the manufacturer. The choice of the absorbent material should be made after carefully considering various factors such as appearance, cost, workability, flame resistance, durability, and light reflection. Following are some of the common types of absorbent materials:

1. Hairfelt: the average value of coefficient of absorption for Hairfelt for 25 mm thickness is 0.60.

2. Acoustic plaster: this is also known as fibrous plaster and it includes granulated insulation material mixed with cement. If quantity of cement is more than required, the plaster will not have sufficient pores to become effective for acoustics. If quantity of cement is less, the plaster will not have enough strength. Thus the quantity of cement should be carefully decided. For thickness of 20 mm and density of 0.10 g per cm^3, the acoustic plaster possesses an absorbent coefficient of 0.30 at 500 cycles per second. Acoustic plaster boards are also available. They can be fixed on the wall and their coefficient of absorption varies from 0.15 to 0.30.

3. Acoustical tiles: these are made in factory and sold under different trade names. The absorption of sound is uniform from tile to tile and they can be fixed easily. However, acoustical tiles are relatively costly than other absorbent

materials. They are most suitable for rooms in which small area is available for acoustical treatment.

4. Strawboard: this material can be used as absorbent material. With a thickness of 13 mm and density of 0.24 g per cm^3, it possesses a coefficient of absorption of 0.30 at 500 cycles per second.

5. Pulp boards: these are soft boards which are prepared from compressed pulp. They are not expensive and can be fixed by ordinary paneling. The average value of coefficient of absorption is 0.17.

6. Compressed fiberboard: this material may be perforated or unperforated. The average coefficient of absorption for perforated board is 0.30 and for the unperforated board is 0.52. It has a density of 0.30 g per cm^3.

7. Compressed wood particle board: this material is provided with perforations and it can be painted also. With a thickness of about 13 mm, the average coefficient of absorption is 0.40.

8. Perforated plywood: this material can be used by forming composite panels with mineral wool and hardboard. It is generally suspended from trusses. The average value of coefficient of absorption for the former composite panel is about 0.20.

9. Wood wool board: this material is generally used with a thickness of 25 mm and it has a density of 0.40 g per cm^3. The average value of coefficient of absorption is 0.20.

10. Quilts and mats: these are prepared from mineral wool or glass wool and are fixed in the form of acoustic blankets. The absorption coefficients of such quilts and mats depend on the thickness, density, perforations, mode of fixing, nature of backing and frequency of sound.

2.2 Noise and its effects

When the sound waves are non-periodic, irregular and of short duration, they produce a displeasing effect and such a sound is known as noise. Thus a noise is an unwanted abrupt sound of complex character with an irregular period and amplitude originating from a source of non-periodic motion.

Following are the important effects of noise: (i) noise creates uncomfortable living conditions; (ii) prolonged exposure to noise may result into temporary deafness or nervous breakdowns; (iii) it is observed that noise has an influence on blood pressure, on muscular strain and even on sleep; (iv) noise leads to fatigue and consequently, the efficiency of persons exposed to noise decreases considerably; (v) it is an established fact that reduction in noise increases to a great extent the output of labor; (vi) presence of noise takes away the essence of music and speech.

2.2.1 Transmission of noise

Any type of noise is transmitted to the room through walls, floors, ceilings or conduits. The origin of transmitted noise may be air-borne or may be due to impact.

Air-borne noise can be transmitted to the receiving room in two ways: (i) by air path between two rooms such as doors, windows, ventilators, key holes, ducts, pipes, and cracks and (ii) by forced vibrations set up by the transmitting room to the walls, floors and ceiling of the receiving room. It is found that air-borne noise sets up forced vibrations in the walls, floors and ceiling of the transmitting room and they in turn set up corresponding vibrations in the walls, floors and ceiling of the receiving room. These surfaces of the receiving room create sound waves and noise is thus transmitted to the receiving room.

Impact noise or structure-borne noise is developed in solid structures and it is then transmitted as air-borne noise. Closing of doors, vibrations of machines, etc., set up vibrations in solid materials of the structure which result in transmission of noise to the receiving room.

Air-borne noise possesses less power, continues for a long duration and is confined to places near its origin. Impact or structure-borne noise possesses more power, continues for a short duration and is often propagated over long distances.

2.3 Integrated sound and thermal insulation

Because some of thermal insulating materials have cellular or open matrix construction, they have inherent ability to absorb sound and act as panel dampers. They also reduce noise breakout, from the machinery plant by their ability to be flexible or discontinuous link between an acoustically active surface and the outer cladding. It should be remembered that sound insulation and sound absorption are quite different terms. The function of a sound insulating construction is to reduce sound passing through it. The function of a sound absorbent material is to reduce sound reflected from a surface. Hence, porous materials in general are good sound absorbers. But they are poor sound insulators. On the other hand, hard materials in general are poor sound absorbers. But they are good sound insulators. Further, insulation of sound is measured in an adjoining room while absorption of sound is measured in the room where sound is produced. The simple material which can be used as insulator is a sheet of material placed in the sound transmission pathways. Sound energy reaches the surface in the form of a pressure wave, of which partial energy passes through the partition and the rest is reflected.

Transmission loss: as the air-borne sound passes through any structure, loss of sound-intensity takes place. This is known as transmission loss.

Important facts: (i) transmission loss is numerically equal to the loss of intensity of noise; (ii) the efficiency of sound and thermo-fluid insulation of a wall or a partition is expressed in terms of transmission loss which occurs when air-borne physical agent passes through the wall or the partition; (iii) transmission loss varies directly with the frequency of physical agent. Hence, transmission loss of a structure should be studied over a wide range of frequencies. (iv) Greater insulation of a wall or a partition is indicated by the larger value of transmission loss.

Methods of sound insulation: the method of sound insulation will depend on the type of noise to be treated and the degree of sound insulation required. The methods of sound insulation can thus be classified in three main categories:

1. When source of noise is in the room itself: following are the methods of sound insulation which are commonly used when the source of noise is situated in the room to be treated for sound insulation:

   a. Improvement in working methods: the basic principle of sound insulation is to suppress the noise at the source itself. A working method creating less noise may be adopted. For instance, the machine in the room is enclosed in a box-like structure with sound absorbing on its surfaces.

b. Acoustical treatment: the walls, floors and ceilings should be provided with sound absorbing materials. The sound absorbing materials should be mounted on the surfaces near the source of noise. The acoustical treatment of the room considerably reduces the noise level in the room.

c. Personal protective devices: it is possible to reduce the noise to some extent by using personal protective devices such as ear plugs and headphones.

2. When noise is air-borne: the sounds generated and transmitted in air directly to human ears are known as air-borne sounds. The air-borne noise possesses less power, continues for a long duration and is confined to places near its origin. Following methods of sound insulation may be adopted for the reduction of air-borne noise:

a. Solid non-porous homogeneous partitions: provision of solid non-porous homogeneous partitions will reduce air-borne noise. It is found that transmission loss of such partitions depends directly on the weight of partition per unit area. The sound insulation of a partition thus increases with the increase in its thickness. But doubling the thickness of a partition reduces transmission loss by a constant amount. This figure is practically constant and can be used to work out the transmission loss of the partition with different thicknesses. It can thus be seen that sound insulation by solid non-porous homogeneous partitions is expensive in quantity of material.

b. Partitions of porous materials: the porous materials may be rigid or flexible. For partitions of rigid porous materials such as concrete masonry, the sound insulation increases about 10% due to the absorptive property of the material. But partitions of flexible porous materials such as wool and quilt do not give enough sound insulation. However, the value of transmission loss decreases as further layers of flexible porous materials are added. The general behavior of partitions of flexible porous materials is such that as the thickness of partition is increased in arithmetic progression, the corresponding transmission loss is in geometric progression. A combination of rigid porous materials and flexible porous materials may be used with advantage for the construction of partition wall. It will provide effective sound insulation and will have less weight.

c. Double wall construction: it is found that a double wall construction is better for sound insulation than a solid wall construction. The walls are of plasterboards or fiberboards or plaster on lath. An air space of about 10 to 12 cm is kept between the walls and staggered wooden studs are provided. In order to make the partition more effective, it is necessary to reduce the number of structural ties between the two parts of the partition to a minimum. The hollow space may be filled with sound absorbing blankets.

d. Floating floor construction: in this type of construction, a floor is separated from the structural floor by means of a layer of resilient material such as mineral and glass wool quilt. Such a floor is known as a floating floor and it results in better sound insulation.

e. Suspended ceiling construction: if a false independent ceiling is constructed below the structural floor, the sound insulation capacity of the floor increases. This construction is useful especially in case of wood-joist floors.

f. Box-type construction: this type of construction gives exceptionally low value of air-borne transmission and hence, it is adopted at places such as broadcasting studios where low air-borne sound transmission is most essential. A box-like structure is construction on the structural floor.

g. Design of doors and windows: for good insulation, it is necessary to design carefully the doors and windows of the room. The sound travels through very thin cracks between the door and wall. The space between the jamb and frame may be packed with sound absorbing material. In case of a door, the transmission loss increases with the increase in weight. In case of a window, the transmission loss increases with increase in thickness of glass. Excellent sound insulation is obtained by constructing glazed windows with double or triple panes of glass. The air space at the edges of such panes is filled with sound absorbing material.

h. Planning of rooms: if rooms within residential buildings are suitably arranged, good sound insulation is achieved. It is also economical than structural measures required for good sound insulation.

3. When noise is structure-borne: the sounds which originate and progress on the building structure are known as structure-borne sounds or impact sounds. The structure-borne noise is powerful, propagates over long distances and persists for a very short duration. Following methods of sound insulation may be adopted for the reduction of structure-borne noise:

a. Treatment of floors and ceilings: the floors and ceilings may be treated for floating floors and suspended ceilings which help in considerably reducing structure-bore sound.

b. Discontinuous construction: this method is similar to box-type construction. The walls of the rooms are constructed on floating floors and the ceilings of the rooms are suspended from the structural floors. The use of structural ties with the main walls is avoided as far as possible or special resilient isolators are employed for this purpose.

c. Insulation of machinery: mechanical equipment such as refrigerators, lifts, and fans create vibrations in the structure and hence, if they are isolated properly, structure-borne sound is reduced to a considerable extent. The main principle of insulation of machinery is to rest the mechanical equipment on a flexible support which may be of rubber, cork, felt or metal spring.

d. Town planning: vibrations from external sources such as railways, metros, cars, traffic, and factories create structure-borne sound. The most effective method for reducing such type of structure-borne sound is to have a rational town planning. The city is divided into suitable zones and residential zone is placed away from railways, workshops, factories and main streets.

2.4 Thermal insulation definition

The temperatures inside and outside a building are different. Some building materials allow heat to pass rapidly while others do not allow passage of heat smoothly. The term thermal resistance is used to indicate the construction by which transmission of heat from or in the room is retarded. The main aim of thermal insulation is to minimize the transfer of heat between outside and inside of the building.

Advantages of thermal insulation: the advantages derived from thermal insulation are as follows: (i) comfort: due to thermal insulation, the room remains cool in summer and warm in winter than outside. Hence, a room provided with thermal insulation gives comfort both in summer and winter. (ii) Fuel saving: due to thermal insulation, transfer of heat from inside to outside of the room is reduced. This results in less quantity of fuel required to maintain the desired temperature in the room. (iii) Condensation: the provision of thermal insulating materials inside a room prevents condensation on interior walls and ceilings. Condensation is the deposition of moisture and it takes place when warm air comes into contact with surfaces having temperature below the dew point. (iv) Water system: the use of thermal insulating materials prevents the freezing of water taps in extreme winter and heat loss in case of hot water system.

General principles: following are the general principles of thermal insulation: (i) the thermal resistance of an insulating material is directly proportional to its thickness. (ii) Provision of an air gap is a very important insulating agent and is very essential. (iii) The thermal resistance of a building depends on its orientation also. The building should be so located that there is maximum transfer of solar energy in winter and there is minimum transfer of solar energy in summer.

Insulating materials: the choice of an insulating material depends on the cost, area to be covered, standard of insulation required and the cost of heating or cooling. The thermal insulating material should be reasonably fire-proof, non-absorbent of moisture, able to resist attack on small insects and not liable to undergo deformation. The usual insulating materials are rockwool, slag wool, fiberboards, flexible blankets, saw dust, wood-shavings, cork board slabs, mineral wool slabs, aluminum foils, products of cement concrete with lightweight aggregates, gypsum board, chip board, gasket cork sheet, foam plastic, etc.

Table 2 shows the density, thermal conductivity and thermal resistivity of some of the common building and insulating materials. In general, it may be stated that the materials of low density provide better thermal insulation than those of higher density.

2.4.1 Thermal insulation of exposed doors and windows

Doors and windows which are exposed transmit heat to a considerable extent. Following methods may be employed to ensure heat insulation of exposed doors and windows: (i) insulating glass or double glass with air space may be provided for glazed doors and windows. This will reduce heat transmission through doors and windows. (ii) In order to reduce incidence of solar heat, projections in the form of sun breakers, weather-sheds, projections, curtains, venetian blinds, etc. may be provided on the exposed doors and windows.

2.4.2 Thermal insulation of exposed roofs

Thermal insulation of exposed roofs may be achieved by treating inside surface or outside surface.

Internal treatment: (i) false ceiling with an air gap may be provided. The ceiling is made of thermal insulating materials. (ii) Light insulating materials may be pasted by suitable adhesives to the inside surfaces of the exposed roofs.

External treatment: (i) suitable shade may be provided on the exposed roof surfaces. (ii) Shinning and reflecting materials may be fixed on the top of exposed roofs. (iii) For flat roofs, an air space may be created by arranging

No.	Material	Density (kg/ m^3)	Thermal conductivity K (kcal/m h °C)	Thermal resistivity (1/K)
1	Artificial stone	1760	1.14	0.88
2	Asbestos cement sheet	1520	0.25	4.00
3	Asphalt	2240	1.05	0.95
4	Cement concrete (1:2:4)	2240 to 2480	1.24	0.80
5	Compressed straw slab	368	0.074	13.51
6	Fiberboard	240 to 400	0.046 to 0.056	21.74 to 17.86
7	Glass	2510	0.905	1.10
8	Glass cellular	168	0.061	16.40
9	Glass fiber	48	0.029	34.50
10	Granite	2640	2.52	0.40
11	Hair felt	80	0.034	29.40
12	Limestone	2180	1.32	0.76
13	Ordinary bricks	1760	0.7 to 1.44	1.43 to 0.70
14	Plastering	1280 to 1600	0.50	2.00
15	Sand-lime bricks	1840	0.93	1.07
16	Sandstone	2000	1.12	0.90
17	Saw dust	192	0.051	19.60
18	Terrazzo	2430	1.363	0.73
19	Timber	480 to 270	0.124	8.00
20	Wood-wool slab	400	0.071	14.08

Table 2.
Density, thermal conductivity and thermal resistivity of some common building and insulating materials.

cement sheets or corrugated galvanized iron sheets on bricks. (iv) Flat roofs may be kept cool by water which may either be stored or sprayed at regular intervals. The surface temperature of the roof is reduced by this method. (v) Thermal insulation of flat roof may also be provided by putting a layer of about 25 mm thickness of coconut pith cement concrete. For this purpose, coconut pith and cement are mixed in dry state and then the mix is transferred to concrete mixer. Water in required quantity is added so as to obtain concrete of workable consistency. It is then conveyed and laid in suitable thickness on the roof. Wet coconut pith may also be used with due care and hard mixing may be adopted for small quantity of work. To avoid loss of water from concrete surface, it is covered by an impermeable layer and then it is allowed to dry in air for a period of 20 days to 1 month. Any cracks which are seen during drying period are filled up by the pith concrete. When the pith concrete layer has fully dried, water-proofing treatment may be given in the usual manner.

2.4.3 Thermal insulation of exposed walls

Following methods may be adopted for thermal insulation of exposed walls: (i) suitable thickness of wall may be provided. (ii) Hollow wall or cavity wall construction may be adopted. (iii) For partitions, an air space may be created by fixing hard boards on battens. (iv) The inside and outside surfaces of exposed wall may be provided with thermal insulating materials in such a way that the value of overall thermal transmittance is brought within desired limit. (v) If it is structurally suitable, the exposed wall may be constructed of thermal insulating materials. (vi) It is found that the application of light-colored whitewash or distemper on the exposed side of the wall will grant substantial thermal insulation.

3. Model assumptions and development

The assumptions used in the development of the model for a building integrated photovoltaic airflow window (BIPV-AW) system as depicted in **Figure 1** are: (i) fully developed heat transfer has been assumed for mixed convection heat transfer assuming a parallel plate wide channel at low air velocities ~0.5 m/s; (ii) temperature variation only along y-axis with lumped temperature distribution along x and z-axes; (iii) applicability of first law of thermodynamics at the surface; (iv) clear sky is applicable; (v) quasi steady state heat transfer analysis has been performed assuming a vertical channel; (vi) uniform average air velocity distribution; (vii) temperature variation only in y-direction (vertical), being taken as lumped in other directions (x-axis and z-axis); (viii) air properties are evaluated at film temperature of 300 K; (ix) negligible heat transfer from side walls/insulation panel and room air zone; (x) conduction (diffusion) equation for performing energy balance on air nodes is not taken into consideration; (xi) negligible thermal storage capacity of duct wall; (xii) no infiltration or air leakage sources from the test section; and (xiii) ambient air and room air temperatures are specified.

The system is discretized into network of two adjacent stacks of control volumes common to surface and air nodes (see **Figure 2**). The energy balances are performed on both surface and air nodes with aid of constitutive relations for noise fields due to solar intensity, sound intensity, airflow power intensity, electric power intensity and heat power intensity. The energy conversion and noise characterization is important, for example, in modeling airflow window and PV solar wall building structures. The resultant noise field due to composite wave elements (of heat, fluid, electricity, sound and sun) is a function of solar irradiation, sound intensity, air gap width, mass flow rate and pressure, wall and air temperatures of the double wall building structure. The integrated noise insulation due to thermal and sound

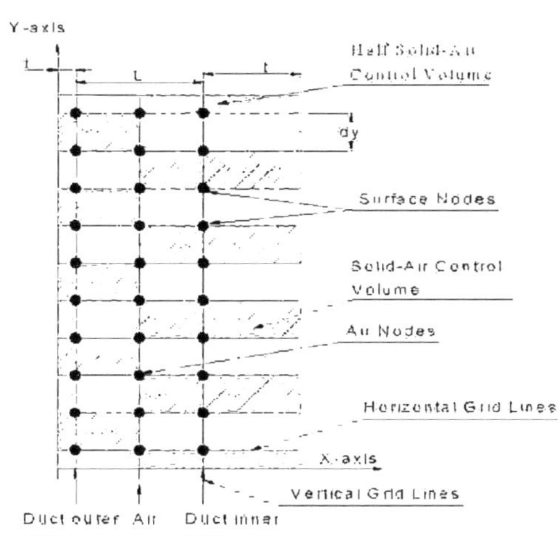

Figure 2.
Grid size in the duct: distribution of nodes and control volumes.

fields is modeled. Noise characterization equations are devised, which calculate noise fields due to ventilation, heat transport and sound transmission of integrated building insulation through a double wall structure.

In writing nodal equations in matrix form, sign notation is adopted for automatic formulation of conductance matrix (U-matrix) with unknown temperatures and heat source elements. The sum of all incoming heat source elements and U-matrix conductances multiplied with temperature difference with respect to the unknown temperatures at other nodes are equal to zero. The energy balance is written in equation form for any general node (m,n):

$$\sum_{n=1}^{N}\left(U_{m,n} \times \Delta T_{m,n}\right)+\sum_{n=1}^{N} Q_{m,n}=0 \tag{1}$$

where $U_{m,n}$ is the conductance at node (m,n), $\Delta T_{m,n}$ is the difference between unknown temperature at the node (m,n) and unknown temperature at surrounding heat exchange node. $Q_{m,n}$ is heat source term at the node (m,n).

3.1 Noise characterization

A unified theory for stresses and oscillations is proposed by Dehra [12]. The following standard measurement equations are derived and adopted from the standard definitions for sources of noise interference [1–60].

Noise of sol: for a pack of solar energy wave, the multiplication of solar power storage and the velocity of light gives solar power intensity I. On taking logarithm of two intensities of solar power, I_1 and I_2, provides intensity difference. It is mathematically expressed as:

$$Sol=\log\left(I_1\right)\left(I_2\right)^{-1} \tag{2}$$

where logarithmic unit ratio for noise of sol is expressed as *Sol*. The oncisol (oS) is more convenient for solar power systems. The mathematical expression by the following equality gives an oncisol (oS), which is 1/11th unit of a *Sol*:

$$oS=\pm 11\,\log\left(I_1\right)\left(I_2\right)^{-1} \tag{3}$$

Noise of therm: for a pack of heat energy wave, the multiplication of total power storage and the velocity of light gives heat power intensity I. The pack of solar energy wave and heat energy wave (for same intensity I), have same energy areas, therefore their units of noise are same as *Sol*.

Noise of photons: for a pack of light energy beam, the multiplication of total power storage and the velocity of light gives light power intensity I. The pack of solar energy wave and light energy beam (for same intensity I), have same energy areas, therefore their units of noise are same as *Sol*.

Noise of electrons: for a pack of electricity wave, the multiplication of total electrical storage and the velocity of light gives electrical power intensity I. The pack of solar energy wave and electricity wave (for same intensity I), have same energy areas, therefore their units of noise are same as *Sol*.

Noise of scattering: for a pack of fluid energy wave, the multiplication of total power storage and the velocity of fluid gives fluid power intensity I. On taking logarithm of two intensities of fluid power, I_1 and I_2, provides intensity difference. It is mathematically expressed as:

$$Sip = \log(I_1)(I_2)^{-1} \tag{4}$$

where logarithmic unit ratio for noise of scattering is *Sip*. The oncisip (oS) is more convenient for fluid power systems.

The mathematical expression by the following equality gives an oncisip (oS), which is 1/11th unit of a *Sip*:

$$oS = \pm 11 \log(I_1)(I_2)^{-1} \tag{5}$$

For energy area determination for a fluid wave, the water with a specific gravity of 1.0 is the standard fluid considered with power of ±1 Wm^{-2} for a reference intensity I_2.

Noise of scattering and lightning: for a pack of fire wave, the intensity, I, of fire flash with power of light, is the multiplication of total power storage and the velocity of light. Whereas for a pack of fire wave, the intensity, I, of fire flash with power of fluid, is the multiplication of total power storage capacity and velocity of fluid.

- For a noise due to fire flash, the collective effect of scattering and lightning is obtained by superimposition principle.
- For same intensity I, the pack of solar energy wave and a fire flash with light power have same energy areas; therefore, their units of noise are same as *Sol*. The therm power may also be included in fire flash with power of light.
- For same intensity I, the pack of fluid energy wave and a fire flash with fluid power have same energy areas; therefore, their units of noise are same as *Sip*. In determining the areas of energy for the case of fluids other than water, a multiplication factor in specific gravity has to be evaluated.

Noise of elasticity: for a pack of sound energy wave, the product of total power storage and the velocity of sound gives sound power intensity I. On taking logarithm of two intensities of sound power, I_1 and I_2, provides intensity difference. It is mathematically expressed as:

$$Bel = \log(I_1)(I_2)^{-1} \tag{6}$$

where logarithmic unit ratio for noise of elasticity is *Bel*. The oncibel (oB) is more convenient for sound power systems. The mathematical expression by the following equality gives an oncibel (oB), which is 1/11th unit of a *Bel*:

$$oB = \pm 11 \log(I_1)(I_2)^{-1} \tag{7}$$

There are following elaborative points on choosing an *onci* as 1/11th unit of noise [37]:

Reference value used for I_2 is –1 W m^{-2} on positive scale of noise and 1 W m^{-2} on negative scale of noise. In a power cycle, all types of wave form one positive power cycle and one negative power cycle [15]. Positive scale of noise has 10 positive units and 1 negative unit, whereas negative scale of noise has 1 positive unit and 10 negative units:

i. Each unit of sol, sip and bel is divided into 11 parts, 1 part is 1/11th unit of noise;

ii. The base of logarithm used in noise measurement equations is 11.

Reference value of I_2 is –1 W m^{-2} with I_1 on positive scale of noise, should be taken with negative noise measurement expression (see Eqs. 3, 5 and 7), therefore it gives positive values of noise;

Reference value of I_2 is 1 W m^{-2} with I_1 on negative scale of noise, should be taken with positive noise measurement expression (see Eqs. 3, 5 and 7), therefore it gives negative values of noise.

The choosing of *onci* in noise units is done so as to have separate market product and system of noise scales and their units distinguished from prevailing *decibel* units (which have its limitations) in the International System of Units. More discussions on energy conversion, noise characterization theory and choice of noise scales and its units are presented in many papers by the author [1–60].

3.2 Noise reduction coefficients

Noise reduction coefficient for noise of sol (thermal power):

$$\mathrm{NRC} = 1 - 11^{-\Delta oS/22} \tag{8}$$

where ΔoS is noise of therm reduction (noise transmission loss) in oncisol.
Noise reduction coefficient for noise of sip (fluid power):

$$\mathrm{NRC} = 1 - 11^{-\Delta oSi/22} \tag{9}$$

where ΔoSi is noise of scattering reduction (noise transmission loss) in oncisip.
Noise reduction coefficient for noise of bel (sound power)

$$\mathrm{NRC} = 1 - 11^{-\Delta oB/22} \tag{10}$$

where ΔoB is noise of elasticity reduction (noise transmission loss) in oncibel.

4. Model assumptions and development

The picture of the experimental setup in a prefabricated outdoor room is presented in **Figure 3**. The solar noon annual solstices and equinoxes days are selected for performing sensitivity analysis to achieve range of: (i) temperatures of pre-conditioned fresh air available; (ii) electric power generation vis-à-vis surface temperatures of photovoltaic modules; (iii) integrated noise insulation values due to noise fields of composite wave elements transmitted into the room. Computer aided simulation model is developed for an airflow window system located in Montréal. Some examples of noise insulation calculations are illustrated using newly devised noise measurement equations for noise of sol, noise of therm, noise of scattering

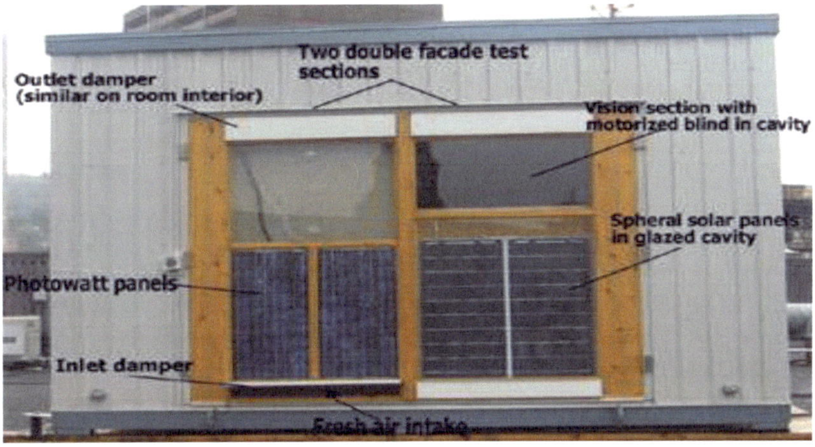

Figure 3.
Prefabricated outdoor room at Concordia University.

and noise of elasticity. The sensitivity analysis for an outdoor duct is also conducted for critical design of ventilation requirements with supply of varying outdoor mass flow rate to a single building zone. The improved method is useful for accurately predicting ventilation air requirements along with designing integrated thermal and sound insulation through a double or cavity wall building structure.

Table 3 has provided properties of physical domain. **Tables 4–10** have presented sensitivity analysis and noise characterization values for the exterior duct based on mass flow rate, solar irradiation and size of duct. The thermal modeling results are presented in **Figures 4–8**. **Figure 4** has presented efficiencies of the building integrated photovoltaic airflow window system viz., electrical efficiency of PV module and combined efficiency of the system. **Figure 5** has presented thermal model results of PV Module, insulation panel and air with respect to height of the spandrel section. **Figure 6** has presented thermal model results for PV module temperatures with solar time for forced and natural convection and air temperatures for forced and natural convection for air cavities I and II. **Figure 7** has results for useful energy generated and solar energy absorbed by a photovoltaic module. **Figure 8** has provided variation of hydraulic diameter, velocity and flow rate vs. pressure drop on a log scale.

The thermo-physical properties of photovoltaic modules, air and insulating panel were assumed constant along all directions i.e. x-, y-, and z-ordinates. The thermo-physical properties of insulating panel with building insulation were obtained from tests conducted with heat flow meter and related specifications from the manufacturer [3]. The temperature differences along x-direction are obtained by assuming same temperature difference per unit thickness of material along x- and y-ordinates [3]. The heat storage capacity for temperature differences across x-direction is negligible of the heat storage capacity for temperature differences across y-direction. Therefore temperatures are assumed uniform and lumped in x-direction. The pair of glass coated photovoltaic modules was having three layers of material viz., a flat sheet of solar cells, with glass face sheets on its exterior and interior sides. The measurements were collected for a pair of successive runs at same solar intensities [3]. The thermal model is validated by comparing its predicted results with those obtained from the experimental apparatus. The agreement between the predictions of the thermal model and experimental results was presented to be very good [3].

Property	Value	Property	Value
Solar irradiation	650 W m^{-2}	Width of air gap	0.025 m
Ambient heat transfer coefficient	13.5 W m^{-2} K^{-1}	Thermal conductivity of air	0.02624 W m^{-1} K^{-1}
Ambient air temperature	-5°C	Specific heat of air (Cp)	1000 J kg^{-1} K^{-1}
Building space temperature	20°C	Density of air	1.1174 kg m^{-3}
Height of duct	3.0 m	Kinematic viscosity of air	15.69×10^{-6} m^2 s^{-1}
Width of duct	1.0 m	Prandtl number of air	0.708
Thickness of outer wall of duct	0.0025 m	Air velocity for obtaining mass flow rate	0.75 m s^{-1}
Absorptance of outer wall with flat black paint	0.95	Stefan Boltzmann constant for surface of duct walls	5.67×10^{-8} W m^{-2} K^{-4}
Thermal conductivity of aluminum alloy for HVAC duct	137 W m^{-1} K^{-1}	Emissivity of back surface of duct walls	0.95
RSI value	1.0 m^2 K W^{-1}	Number of nodes in x direction	Nx = 3
Thickness	0.04 m	Number of nodes in y direction	Ny = 10, Δy = 0.3 m

Table 3.
Properties of physical domain.

Solar irradiation (Wm^{-2})	Air temperature difference (ΔT) °C	Noise of sol oS (oncisol)
450	15.50	28
550	18.90	28.93
650	22.40	29.7
750	25.90	30.36
850	29.40	30.91

Table 4.
Temperature difference and noise of sol with solar irradiation (air velocity: 0.75 ms^{-1}).

Air velocity (ms^{-1})	Fluid power (Wm^{-2})	Air temperature difference (ΔT) °C	Noise of scattering oS (oncisip)
1.35	47.62	15.28	17.72
1.05	37.0	18.22	16.50
0.75	26.45	22.40	15.02
0.45	15.87	28.15	12.65
0.15	05.29	29.80	07.64

Table 5.
Temperature difference and noise of scattering with air velocity (S = 650 Wm^{-2}).

(ΔT) °C	Mass flow rate ($kg\,s^{-1}$)	Thermal power (Wm^{-2})	Noise of therm oS (oncisol)	(ΔT) °C	Mass flow rate ($kg\,s^{-1}$)	Thermal power (Wm^{-2})	Noise of therm oS (oncisol)
15.50	0.01376	71.09	19.5602	15.28	0.0231	117.65	21.868
18.90	0.01275	80.325	20.119	18.22	0.0171	103.85	21.296
22.40	0.0120	89.6	20.614	22.40	0.0120	89.6	20.614
25.90	0.0115	99.2833	21.043	28.15	8.1×10^{-3}	76.0	19.866
29.40	0.0111	108.78	21.505	29.80	6.2×10^{-3}	61.59	18.898

Table 6.
Mass flow rate and noise of therm with (ΔT) °C.

Air velocity ($m{\cdot}s^{-1}$)	Fluid power ($W{\cdot}m^{-2}$)	Noise of scattering oS (oncisip)	Sound pressure ($N{\cdot}m^{-2}$)	Sound power intensity ($W{\cdot}m^{-2}$)	Noise of elasticity oB (oncibel)
1.35	47.62	17.72	557.5	752.7	30.36
1.05	37.0	16.50	433.65	455.33	28.05
0.75	26.45	15.02	309.75	232.31	24.97
0.45	15.87	12.65	185.85	83.63	20.24
0.15	05.29	07.64	61.94	09.29	10.12

Table 7.
Noise of elasticity with air particle velocity (impedance $Z_0 = 413\ N{\cdot}s{\cdot}m^{-3}$ at 20°C).

Noise of therm oS (oncisol)—duct	(ΔT) °C—room	Mass flow rate (kg s^{-1})—room	Noise of therm oS (oncisol)—room	Transmission loss ΔoS (oncisol)	Noise reduction coefficient
21.868	25	0.02095	28.721	6.853	0.52619

Table 8.
Thermo-fluid noise transmission loss and noise reduction coefficient.

Noise of elasticity oB (oncibel)—duct $v = 0.75\ m{\cdot}s^{-1}$	Noise of elasticity oB (oncibel)—room $v = 0.15\ m{\cdot}s^{-1}$	Transmission loss ΔoB (oncibel)	Noise reduction coefficient
24.97	9.29	15.68	0.81896

Table 9.
Acoustic noise transmission loss and noise reduction coefficient.

Noise of scattering oS (oncisip)—duct $v = 0.75\ m{\cdot}s^{-1}$	Noise of scattering oS (oncisip)—room $v = 0.15\ m{\cdot}s^{-1}$	Transmission loss ΔoS (oncisip)	Noise reduction coefficient
15.02	7.64	7.38	0.55264

Table 10.
Fluid noise transmission loss and noise reduction coefficient.

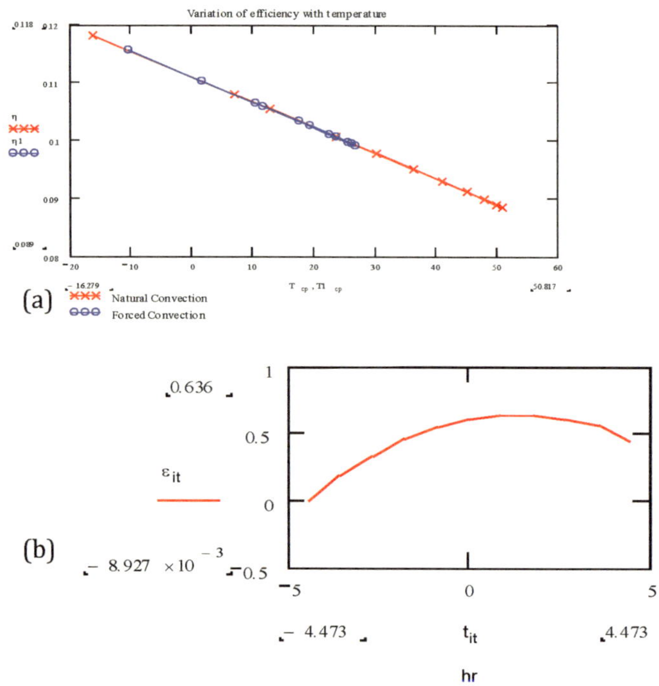

Figure 4.
Efficiencies: (a) electrical efficiency of PV module and (b) combined efficiency of the system.

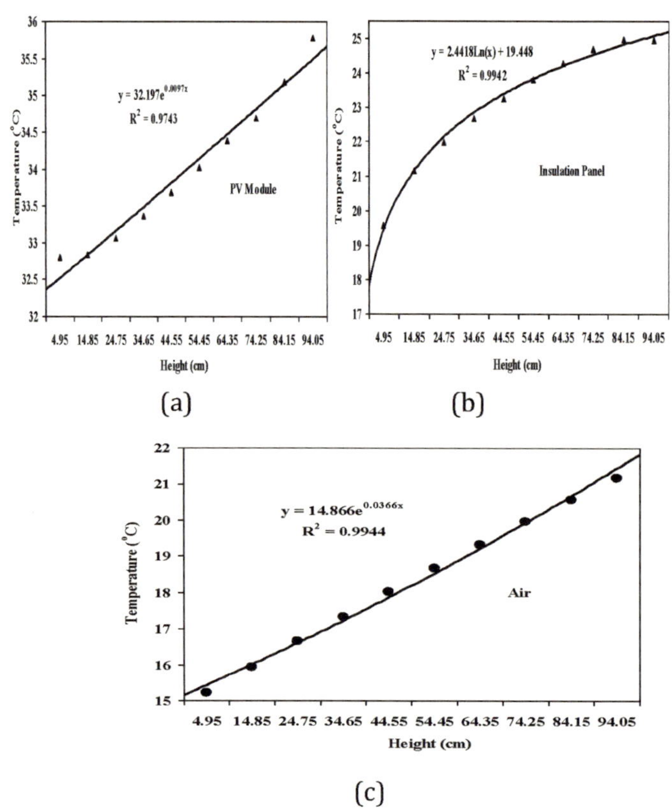

Figure 5.
Thermal model results: (a) PV module, (b) insulation panel, and (c) air.

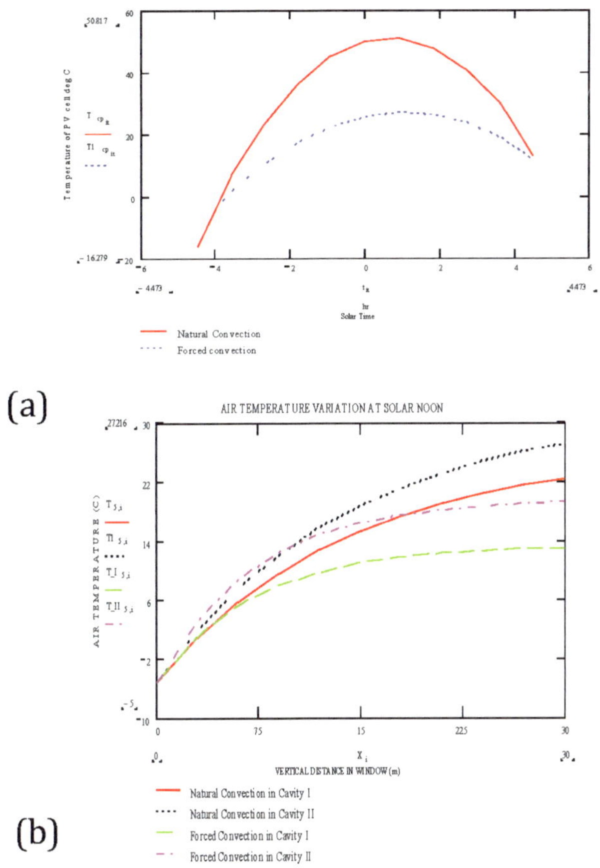

Figure 6.
Thermal model results: (a) PV module temperatures and (b) air temperatures.

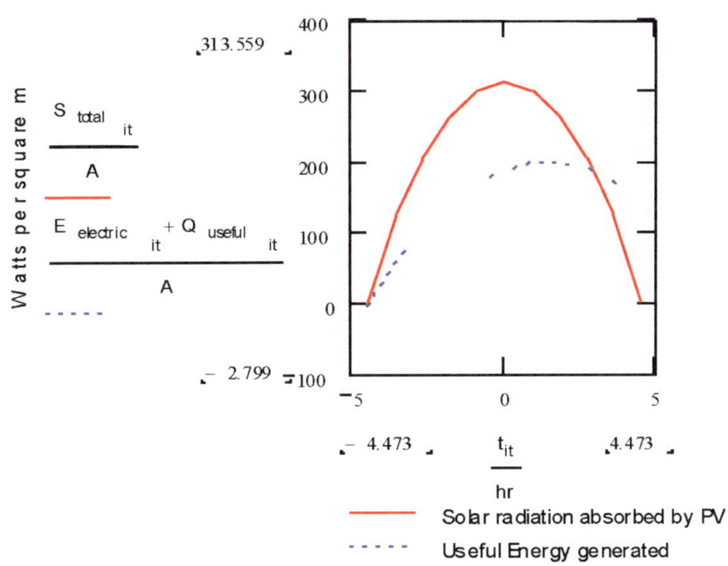

Figure 7.
Useful energy generated and solar energy absorbed by a photovoltaic module.

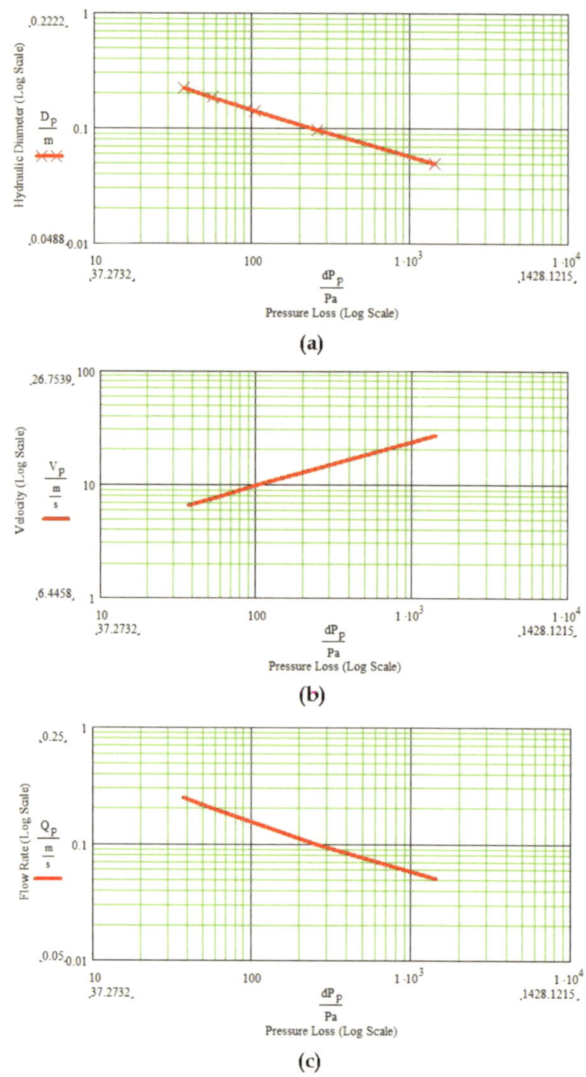

Figure 8.
Variation of (a) hydraulic diameter, (b) velocity, and (c) flow rate vs. pressure drop on a log scale.

5. Conclusion

A study on integrated insulation modeling of an airflow window with a PV solar wall via energy conversion is performed. The noise interference and characterization equations as per speed of a composite wave are presented. The sources of noise measurement equations (sun, light, sound, heat, electricity, fluid and fire) are described depending on their speed of noise interference. Noise measurement equations and their units are coined. The acoustic insulation systems are classified as per source signals of solar power, electric power, light power, sound power, heat power, fluid power and fire power. Based on sensitivity analysis conducted on an outdoor duct exposed to solar radiation, integrated insulation model has calculated noise transmission losses and noise reduction coefficients, besides calculating individual values of oncisol, oncisip and oncibel for various types of noises.

Several performance and optimization issues are considered in development of the model including optimal air velocity for heat transfer, dimensions of PV module (height), selection of cavity width to reduce pressure drops, and prediction of temperature rise of air as it flows out of the airflow window system and into the outdoor test-room. The airflow is adjusted to a constant value to optimize necessary temperature for integrated photovoltaic array as well as for pre-heated fresh air into the outdoor test-room. It is envisaged that inside an airflow window integrated with PV, cooling by forced convection is essential, without which, the temperature of PV cell reaches very high (51°C), which decreases the efficiency by more than 20% [4]. The combined efficiency (electrical and thermal) of the system reaches 50%.

A building integrated photovoltaic airflow window (BIPV-AW) system is developed for the purpose of combined generation of electricity, thermal energy and daylighting. This approach will have additional following advantages: (a) there will be reduction in peak heating loads, which will reduce the required capacity of the heating/cooling system; (b) there will be reduction in energy consumed for heating and lighting in the building; and (c) electricity demand of the building will be reduced and energy utilities will get peak surplus.

Acknowledgements

The author conducted part of the work at Department of Building, Civil and Environmental Engineering, Concordia University, Montréal, Québec, Canada.

Author details

Himanshu Dehra
Monarchy of Concordia, India

*Address all correspondence to: anshu_dehra@hotmail.com

References

[1] Asdrubali F, D'Alessandro F, Schiavoni S. A review of unconventional sustainable building insulation materials. Sustainable Materials and Technologies. 2015;**4**:1-17

[2] Asdrubali F, Schiavoni S, Horoshenkov KV. Review of sustainable materials for acoustic applications. Building Acoustics. 2012;**19**(4):283-312

[3] Dehra H. A numerical and experimental study for generation of electric and thermal power with photovoltaic modules embedded in building façade [submitted/unpublished PhD dissertation]. Montréal, Québec, Canada: Department of Building, Civil and Environmental Engineering, Concordia University; 30 August 2004. p. 120

[4] Dehra H. Modelling of airflow window with integrated photovoltaics [doctoral seminar]. Montréal, Canada: Department of Building, Civil and Environmental Engineering, Concordia University; 2002

[5] Dehra H. A 1-D/2-D model for an exterior HVAC rectangular duct with a steady solar heat flux generation. In: Orally Presented at the Proceedings of the Second International Green Energy Conference; 25-29 June 2006; Oshawa, Canada. 2006. pp. 1240-1251

[6] Dehra H. The effect of heat and thermal storage capacities of photovoltaic duct wall on co-generation of electric and thermal power. In: Orally Presented in the Proceedings of American Institute of Chemical Engineers - AIChE 2007 Spring Meeting; 22-26 April 2007; Houston, Texas, USA; Session 36a. 2007 (9 pages). Available from: https://www3.aiche.org/proceedings/Abstract.aspx?PaperID=79681

[7] Dehra H. On solar building energy devices. In: Orally Presented at the 18th IASTED International Conference on Modelling and Simulation; 30 May-1 June 2007; Montréal, Canada. 2007. pp. 96-101. Available from: https://www.iasted.org/conferences/servepdf-567.html?name=FinalProgram.pdf; http://dl.acm.org/citation.cfm?id=1295663

[8] Dehra H. A heat transmission model for a telephone line. In: Orally Presented at the 21st Canadian Congress of Applied Mechanics, Department of Mechanical & Industrial Engineering, Ryerson University; 3-7 June 2007; Toronto, Ontario, Canada. 2007. pp. 356-357. Available from: http://www.worldcat.org/title/cancam-2007-21st-canadian-congress-of-applied-mechanics-ryerson-university-toronto-ontario-canada-june-3-7-2007-proceedings-le-21ieme-congres-canadien-de-mecanique-appliquee-comptes-rendus/oclc/500984841

[9] Dehra H. The study of a solar cell device under illumination. In: Orally Presented at the 21st Canadian Congress of Applied Mechanics, Department of Mechanical & Industrial Engineering, Ryerson University; 3-7 June 2007; Toronto, Ontario, Canada. 2007. pp. 354-355. Available from: http://www.worldcat.org/title/cancam-2007-21st-canadian-congress-of-applied-mechanics-ryerson-university-toronto-ontario-canada-june-3-7-2007-proceedings-le-21ieme-congres-canadien-de-mecanique-appliquee-comptes-rendus/oclc/500984841

[10] Dehra H. Thermal modelling of an airflow window with a photovoltaic solar wall. In: Accepted Abstract in the Proceedings of The 3rd International Energy, Exergy and Environment Symposium; 1-5 July 2007; Évora, Portugal. 2007 (1 page)

[11] Dehra H. The electrodynamics of a pair of PV modules with connected building resistance. In: Presented at The 3rd IASME/WSEAS International Conference on Energy, Environment, Ecosystems and Sustainable Development (EEESD'07); 24-26 July 2007; Agios Nikolaos, Crete Island, Greece. 2007. pp. 563-567. Available from: http://www.wseas.us/e-library/conferences/2007creteeeesd/papers/562-125.pdf

[12] Dehra H. A unified theory for stresses and oscillations. In: Orally Presented at the Canadian Acoustical Association (CAA) Montréal 2007 Conference, Concordia University; 9-12 October 2007; Montréal, Canada. 2007. pp. 132-133. Available from: https://jcaa.caa-aca.ca/index.php/jcaa/article/view/1936

[13] Dehra H. A model for stresses and oscillations of a resonant cavity wall. In: Accepted Abstract for an Oral Presentation at 57th Canadian Chemical Engineering Conference; 28-31 October 2007; Edmonton, Alberta, Canada. 2007 (1 page). Available from: http://csche2007.cheminst.ca/00000039.htm

[14] Dehra H. Mathematical analysis of a solar thermosyphon. In: Accepted for an Oral Presentation at the 1st International Conference on Advances in Energy Research, Indian Institute of Technology (IIT) Bombay; 12-14 December 2007, Mumbai, India. 2007 (6 pages). Available from: http://www.printsasia.com/book/international-conference-on-advances-in-energy-research-icaer-2007-proceedings-of-the-first-international-023063432x-9780230634329

[15] Dehra H. Power transfer and inductance in a star connected 3-phase RC circuit amplifier. In: Proceedings of American Institute of Chemical Engineers - AIChE 2008 Spring Meeting; 6-10 April 2008; New Orleans, LA, USA; Session 96a. 2008 (7 pages). Available from: https://aiche.confex.com/aiche/s08/techprogram/P103422.HTM

[16] Dehra H. The representation of units on hexagonal scales of noise. In: Accepted Abstract in the Proceedings of the 58th Canadian Chemical Engineering Conference; 19-22 October 2008; Ottawa, Ontario, Canada. 2008 (1 page). Available from: http://csche2008.cheminst.ca/00001032.htm

[17] Dehra H. A benchmark solution for interference of noise waves. In: AIChE Spring 2009; 26-30 April 2009; Tampa, FL, USA; Session 67c. 2009 (4 pages). Available from: https://aiche.confex.com/aiche/s09/webprogram/Paper143116.html

[18] Dehra H. A guide for signal processing of sensors and transducers. In: AIChE Spring 2009; 26-30 April 2009; Tampa, FL, USA; Session 6b. 2009 (9 pages). Available from: https://aiche.confex.com/aiche/s09/webprogram/Paper143251.html

[19] Dehra H. Acoustic filters. In: Proceedings National Conference on Renewable Energy (NaCORE 2009); 5-7 November 2009; Jodhpur, Rajasthan, India. 2009. pp. 164-168. Available from: http://www.dkagencies.com/doc/from/1123/to/1123/bkId/DK9133217162767395251117561/details.html

[20] Dehra H. Ventilation materials characterisation of a solar panel duct wall. In: Proceedings National Conference on Renewable Energy (NaCORE 2009); 5-7 November 2009; Jodhpur, Rajasthan, India. 2009. pp. 235-241. Available from: http://www.dkagencies.com/doc/from/1123/to/1123/bkId/DK9133217162767395251117561/details.html

[21] Dehra H. A building integrated photovoltaic airflow window system. In: 2nd International Conference on Advances in Energy Research, Indian

Institute of Technology (IIT) Bombay, Mumbai, India, December 09-11, 2009 (6 pages)

[22] Dehra H. Experiments on photovoltaic modules embedded in building façade. In: AIChE Spring 2010; 21-25 March 2010; San Antonio, TX, USA; Session 38b. 2010 (9 pages). Available from: https://www3.aiche.org/proceedings/Abstract.aspx?PaperID=171997

[23] Dehra H. A monitoring model for a fan operated photovoltaic hybrid air ventilation system. In: AIChE Spring 2010; 21-25 March 2010; San Antonio, TX, USA; Session 143b. 2010 (11 pages). Available from: https://www.aiche.org/conferences/aiche-spring-meeting-and-global-congress-on-process-safety/2010/proceeding/paper/143b-monitoring-model-fan-operated-photovoltaic-hybrid-air-ventilation-system-0

[24] Dehra H. A design tool for an outdoor duct. In: Proceedings 9th International Conference on Sustainable Energy Technologies (SET 2010), Shanghai Jiao Tong University School of Medicine; 24-27 August 2010; Shanghai, China. 2010 (7 pages). Available from: http://www.worldcat.org/title/9th-international-conference-on-sustainable-energy-technologies-set-2010/oclc/729681678

[25] Dehra H. A portable solar farm house. In: AIChE Spring 2011; 13-17 March, 2011; Chicago, IL, USA; Session 116d. 2011 (9 pages). Available from: https://www3.aiche.org/proceedings/Abstract.aspx?PaperID=210477

[26] Dehra H. A slide rule for noise measurement. In: 10th International Conference on Sustainable Energy Technologies (SET 2011); 4-7 September 2011; Istanbul, Turkey. 2011 (5 pages). Available from: https://www.scribd.com/document/106892829/Programme-Final

[27] Dehra H. Modelling of an airflow window with a photovoltaic solar wall. In: 10th International Conference on Sustainable Energy Technologies (SET 2011); 4-7 September 2011; Istanbul, Turkey. 2011 (6 pages). Available from: https://www.scribd.com/document/106892829/Programme-Final

[28] Dehra H. Numerical simulations and experiments on a photovoltaic solar wall. In: Accepted Abstract in AIChE Spring 2012; 1-5 April 2012; Houston, TX, USA; Session 91b. 2012 (1 page). Available from: https://www.aiche.org/conferences/aiche-spring-meeting-and-global-congress-on-process-safety/2012/proceeding/paper/91b-numerical-simulations-and-experiments-on-photovoltaic-solar-wall-0

[29] Dehra H. Photovoltaic solar wall: 2-D numerical modeling and experimental testing under fan induced hybrid ventilation. In: Orally Presented at the IEEE International Conference on Energy Efficient Technologies for Sustainability; St. Xavier's Catholic College of Engineering, Nagercoil, Kanyakumari District, Tamil Nadu, India, IEEE Xplore; 7-8 April 2016. pp. 668-676. Available from: http://ieeexplore.ieee.org/document/7583836/

[30] Dehra H. A novel theory of psychoacoustics on noise sources, noise measurements and noise filters. In: Orally Presented at the NoiseCon16; 13-15 June 2016; Providence, Rhode Island. USA: Institute of Noise Control Engineering; 2016. pp. 933-942. Available from: http://www.ingentaconnect.com/contentone/ince/incecp/2016/00000252/00000001/art00008?crawler=true

[31] Dehra H. On sources and measurement units of noise.

In: Orally Presented at the International Conference on Innovation, Management and Industrial Engineering (IMIE 2016); 5-7 August 2016; Kurume, Fukuoka, Japan. 2016. pp. 219-227. Available from: http://iainst.org/imie2016/

[32] Dehra H. A paradigm for characterization and checking of a human noise behavior. In: WASET Proceedings of 19th International Conference on Psychological and Brain Sciences; 11-12 May 2017; Montréal, Canada. 2017. pp. 658-666. Available from: https://www.waset.org/downloads/17/presentations/17ca050361.pdf; https://www.waset.org/abstracts/70655

[33] Dehra H. Electrical and thermal characteristics of a photovoltaic solar wall with passive and active ventilation through a room. In: WASET Proceedings of 19th International Conference on Sustainable and Renewable Energy Engineering; 11-12 May 2017; Montréal, Canada. 2017. pp. 584-592. Available from: https://www.waset.org/downloads/17/presentations/17ca050297.pdf

[34] Dehra H. A multi-parametric PV solar wall device. In: Proceedings of IEEE International Conference on Power, Control, Signals and Instrumentation Engineering (ICPCSI-2017); 21-22 September 2017; Chennai, India. IEEE; 2017. pp. 392-401. Available from: https://ieeexplore.ieee.org/document/8392324/

[35] Dehra H. Characterization of noise in power systems. In: Orally Presented at IEEE International Conference on Power Energy, Environment & Intelligent Control (PEEIC2018); 13-14 April 2018; Greater Noida, India. IEEE; 2018. pp. 321-330. Available from: http://www.glbitm.org/PEEIC/index.html

[36] Dehra H. A paradigm of noise interference in a wave. In: Internoise-2018, Orally Presented at 47th International Congress and Exposition on Noise Control Engineering; 26-29 August 2018; Chicago, Illinois, USA. 2018. pp. 451-462. Available from: https://internoise2018.org/

[37] Dehra H. Acoustic signal processing and noise characterization theory via energy conversion in a PV solar wall device with ventilation through a room. Advances in Science, Technology and Engineering Systems Journal. 2018;**3**(4):130-172

[38] Dehra H. The entropy matrix generated exergy model for a photovoltaic heat exchanger under critical operating conditions. International Journal of Exergy. 2008;**5**(2):132-149. DOI: 10.1504/IJEX.2008.016671. Available from: http://www.inderscience.com/info/inarticle.php?artid=16671

[39] Dehra H. The noise scales and their units. Canadian Acoustics. 2008;**36**(3):78-79

[40] Dehra H. A two dimensional thermal network model for a photovoltaic solar wall. Solar Energy. 2009;**83**(11):1933-1942. DOI: 10.1016/j.solener.2008.07.014. Available from: http://www.sciencedirect.com/science/article/pii/S0038092X08001862

[41] Dehra H. A combined solar photovoltaic distributed energy source appliance. Natural Resources. 2011;**2**(2):75-86. DOI: 10.4236/nr.2011.22010. Available from: http://www.scirp.org/journal/PaperInformation.aspx?PaperID=5258

[42] Dehra H. A theory of acoustics in solar energy. Natural Resources. 2013;**4**(1A):116-120. DOI: 10.4236/nr.2013.41A014. Available from: http://www.scirp.org/journal/PaperInformation.aspx?PaperID=29361

[43] Dehra H. A mathematical model of a solar air thermosyphon integrated with building envelope. International Journal of Thermal Sciences. 2016;**102**:210-227. DOI: 10.1016/j.ijthermalsci.2015.11.022. Available from: http://www.sciencedirect.com/science/article/pii/S1290072915003750

[44] Dehra H. Photovoltaic solar wall: 2-D numerical modeling and experimental testing under fan induced hybrid ventilation. New York, USA: IEEE Xplore. 2016. pp. 668-676. Available from: http://ieeexplore.ieee.org/document/7583836/

[45] Dehra H. An investigation on energy performance assessment of a photovoltaic solar wall under buoyancy-induced and fan-assisted ventilation system. Amsterdam, Netherlands: Applied Energy. 2017;**191**:55-74. DOI: 10.1016/j.apenergy.2017.01.038. Available from: http://www.sciencedirect.com/science/article/pii/S0306261917300466

[46] Dehra H. Acoustic signal processing and noise characterization theory via energy conversion in a PV solar wall device with ventilation through a room. Advances in Science, Technology and Engineering Systems Journal. 2018;**3**(4):130-172. DOI: 10.25046/aj030414. Available from: https://www.astesj.com/publications/ASTESJ_030414.pdf; https://astesj.com/v03/i04/p14/

[47] Dehra H. Acoustic filters for sensors and transducers, ICAE2018. Energy Procedia. 2019;**158**:4023-4030. DOI: 10.1016/j.egypro.2019.01.837

[48] Dehra H. Modeling of energy conversion and noise characterization in outdoor ducts exposed to solar radiation, ICAE2018. Energy Procedia. 2019;**158**:5162-5168. DOI: 10.1016/j.egypro.2019.01.681. Available from: https://www.sciencedirect.com/science/article/pii/S1876610219307167/pdfft?md5=25603d1f35661f3b43beb6e295763507&pid=1-s2.0-S1876610219307167-main.pdf

[49] Dehra H. Principles of energy conversion and noise characterization in air ventilation ducts exposed to solar radiation, special section: ICAE2018. Applied Energy. 2019;**242C**:1320-1345. DOI: 10.1016/j.apenergy.2019.03.013. Available from: https://www.sciencedirect.com/science/article/pii/S0306261919304040

[50] Dehra H. Characterization of noise in power systems. New York, USA: IEEE Xplore. IEEE; 2019. pp. 320-329. DOI: 10.1109/PEEIC.2018.8665443. Available from: https://ieeexplore.ieee.org/document/8665443

[51] Dehra H. Monarchy of Concordia: A globalized society on maintaining peace and harmony in the world by controlling human noise behavior. International Journal of Social Sciences. 2020;**9**(1). DOI: 10.20472/SS2020.9.1.001. Available from: www.iises.net; https://www.iises.net/international-journal-of-social-sciences/publication-detail-25601; https://www.iises.net/international-journal-of-social-sciences/publication-detail-25601?download=1

[52] Dehra H. Data for Monarchy of Concordia: A Globalized Society on Maintaining Peace and Harmony in the World by Controlling Human Noise Behavior (Version v1) [Data Set]. Geneva, Switzerland: Zenodo; 2019. DOI: 10.5281/zenodo.3515734

[53] Dehra H. Solar energy absorbers. In: Manyala R, editor. Solar Collectors and Panels, Theory and Applications. London, UK: InTechOpen; 2010. pp. 111-134. DOI: 10.5772/10334. Available from: https://www.intechopen.com/books/solar-collectors-and-panels--theory-and-applications/solar-energy-absorbers

[54] Dehra H. Acoustic filters. In: Romano VA, Duval AS, editors.

Ventilation: Types, Standards and Problems. New York, USA: Nova Publishers; 2012. pp. 135-154. Available from: https://www.novapublishers.com/catalog/product_info.php?products_id=40124&osCsid=a5b966ac2674b1639843fc0b6d201172

[55] Dehra H. Solar thermosyphon. In: Méndez-Vilas A, editor. Materials and Processes for Energy: Communicating Current Research and Technological Developments, Formatex Energy Book Series. 2013. pp. 142-149

[56] Dehra H. Chapter 11: A theory of acoustics in solar energy. In: Keach S, editor. Recent Progress in Natural Resources. New York, USA. Vol. 2. Academic Pages; 2017. pp. 77-81

[57] Dehra H. Chapter 15: Building-integrated thermoelectric cooling-photovoltaic (TEC-PV) devices. In: Aranguren P, editor. Bringing Thermoelectricity into Reality. London, UK: IntechOpen; 2018. pp. 313-330. DOI: 10.5772/intechopen.75472. Available from: https://www.intechopen.com/books/bringing-thermoelectricity-into-reality/building-integrated-thermoelectric-cooling-photovoltaic-tec-pv-devices

[58] Dehra H. Chapter 1: Solar energy conversion and noise characterization in photovoltaic devices with ventilation. In: Prabaharan N, Rosen MA, Campana PE, editors. Recent Developments in Photovoltaic Materials and Devices. London, UK: IntechOpen; 2019. pp. 1-20. DOI: 10.5772/intechopen.79706. Available from: https://www.intechopen.com/chapter/pdf-download/62929

[59] Dehra H. Acoustic filters for sensors and transducers: Energy policy instrument for monitoring and evaluating holy places and their habitants. In: Taner T, editor. Energy Policy. London, UK: IntechOpen; 2019. DOI: 10.5772/intechopen.81949. Available from: https://www.intechopen.com/chapter/pdf-download/64590

[60] Dehra H. Noise calculation charts and indoor environmental quality for evaluating industrial indoor environment and health. In: Korhan O, editor. Indoor Environment and Health. London, UK: IntechOpen; 2019. DOI: 10.5772/intechopen.84993. Available from: https://www.intechopen.com/chapter/pdf-download/67373

Chapter 7

A New BEM for Modeling of Acoustic Wave Propagation in Three-Temperature Nonlinear Generalized Magneto-Thermoelastic ISMFGA Structures Using Laser Ultrasonics

Mohamed Abdelsabour Fahmy

Abstract

The principal aim of this chapter is to introduce a new theory called acoustic wave propagation of three-temperature nonlinear generalized magneto-thermoelasticity, and we propose a new boundary element model for solving problems of initially stressed multilayered functionally graded anisotropic (ISMFGA) structures using laser ultrasonics, which connected with the proposed theory. Since there are no available analytical or numerical solutions for the considered nonlinear wave propagation problems in the literature, we propose a new boundary element modeling formulation for the solution of such problems. The numerical results are depicted graphically to show the propagation of three temperatures and displacement waves. The results also show the effects of initial stress and functionally graded material on the displacement waves and confirm the validity and accuracy of our proposed theory and solution technique.

Keywords: boundary element method, acoustic wave propagation, three-temperature, nonlinear generalized magneto-thermoelasticity, initially stressed multilayered functionally graded anisotropic structures, laser ultrasonics

1. Introduction

Physically, according to particle motion orientation and energy direction, there are three wave types, which are categorized as mechanical waves, electromagnetic waves, and matter waves. Mechanical waves are waves, which cannot travel through a vacuum and can travel through any medium at a wave speed, which depends on elasticity and inertia. There are three types of mechanical waves: longitudinal, transverse, and surface waves. Longitudinal waves occur when the movement of the particles is parallel to the energy motion like sound waves and pressure waves. Transverse waves appear when the movement of the particles is perpendicular to the energy motion like light waves, polarized waves, and electromagnetic waves. Surface waves happen when the movement of the particles is in a circular motion. These waves usually occur at interfaces like ocean waves and cup of water ripples. Electromagnetic waves are generated by a fusion of electric and magnetic fields. These waves travel through a vacuum and do not need a medium to travel like microwaves, X-ray, radio waves, and ultraviolet waves. The matter has a wave–particle duality property, where in 1905, Albert Einstein introduced a quantum mechanics theory stating that light has a dual nature; when the light is moving, it shows the wave properties, and when it is at rest, it shows the particle properties, where each light particle has an energy quantum called a photon. Sound is a pressure variation, where a condensation is an increased pressure region on a sound wave and a dilation is a decreased pressure region on a sound wave. Acoustics is the science of study related to the study of sound in gases, liquids, and solids including subjects such as vibration, sound, ultrasound, and infrasound and has grown to encompass the realm of ultrasonics and infrasonics in addition to the audio range, as the result of applications in oceanology, materials science, medicine, dentistry, communications, industrial processes, petroleum and mineral prospecting, music and voice synthesis, marine navigation, animal bioacoustics, and noise cancelation. There are two mechanisms that have been proposed to explain wave generation, which depend on the energy density of laser pulse, a first mechanism at high-energy density, where a thin layer of solid material melts, followed by a dissolution process where the particles fly off the surface, which leads to forces that generate ultrasound, and a second mechanism at low-energy density, where irradiation of laser pulses onto a material generates elastic waves due to the thermoelastic process of expansion of a surface at a high rate. Ultrasound generation with lasers offers a number of advantages over conventional generation with piezoelectric transducers. Since the ultrasound generation by a laser pulse in the thermoelastic range does not damage the material surface, it has several applications such as fiber-optic communication, narrow-band and broadband systems, the ability to work on hard to reach places, curved and rough surfaces, absolute beam energy measurements, and digital images having higher spatial resolution. The process of converting a laser source into an equivalent set of stress boundary conditions takes the largest share of the effort involved in modeling of laser-generated ultrasound, which is very useful in describing the features of a laser-generated ultrasonic in the thermoelastic system [1–3]. Due to the interaction between laser light and a metal surface, the generation of high-frequency acoustic pulses causes the laser irradiation of a metal surface. It led to great progress to develop theoretical models to describe the experimental data [4]. Scruby et al. [5] demonstrated that the thermoelastic area source has been reduced to a point-source influential on the surface. This source point ignores the optical absorption of laser energy into the bulk material and the thermal diffusion from the heat source. Moreover, it does not take into account the limited side dimensions of the source. Rose [6] introduced surface center of expansion (SCOE) based on point-source representation. The SCOE models predict the major features of laser-generated ultrasound waves and agree with experiments particularly well for highly focused Q-switched laser pulses. It fails to predict a precursor in ultrasonic waveforms on and near the epicenter. The precursor is a small sharp initial spike observed in metals signaling the arrival of the longitudinal wave. Doyle [7] established that the existence of the metal precursor is due to subsurface sources which arise from thermal diffusion, since the optical absorption depth is very small in comparison to the thermal diffusion length. According to McDonald [8], Spicer [9] used the generalized thermoelasticity theory to constitute a real model, taking into consideration spatial–temporal shape of the laser pulse and the effect of thermal diffusion.

The mathematical foundations of three-temperature thermoelasticity were defined for the first time by Fahmy [10–14]. Analytical solutions for the current nonlinear generalized thermoelastic problems which are associated with the proposed theory are very difficult to obtain, so many numerical methods were developed for solving such problems like finite difference method [15], discontinuous Galerkin method [16], finite element method (FEM) [17], boundary element method (BEM) [18–31], and other developed techniques [32–36]. The boundary element method [37–67] is actualized effectively for tackling a few designing and logical applications because of its straightforwardness, precision, and simplicity of execution.

In the present chapter, we introduce a new acoustic wave propagation theory called three-temperature nonlinear generalized magneto-thermoelasticity, and we propose a new boundary element technique for modeling problems of initially stressed multilayered functionally graded anisotropic (ISMFGA) structures using laser ultrasonics, which connected with the proposed theory, where we used the three-temperature (3T) radiative heat conduction equations combined with electron, ion, and photon temperatures in the formulation of such problems. The numerical results are presented graphically to show the effects of three temperatures on the displacement wave propagation in the x-axis direction of ISMFGA structures. The numerical results also show the propagation of the displacement waves of homogenous and functionally graded structures under the effect of initial stress. The validity and accuracy of our proposed model was demonstrated by comparing our BEM results with the corresponding FDM and FEM results.

A brief summary of the paper is as follows: Section 1 introduces the background and provides the readers with the necessary information to books and articles for a better understanding of wave propagation problems in three-temperature nonlinear generalized magneto-thermoelastic ISMFGA structures and their applications. Section 2 describes the formulation of the new theory and introduces the partial differential equations that govern its related problems. Section 3 outlines continuity and initial and boundary conditions of the considered problem. Section 4 discusses the implementation of the new BEM and its implementation for solving the governing equations of the problem to obtain the three temperatures and displacement fields. Section 5 presents the new numerical results that describe the displacement waves and three-temperature waves under the effect of initial stress on the homogeneous and functionally graded structures.

2. Formulation of the problem

Consider a multilayered structure with n functionally graded layers in the yz-plane of a Cartesian coordinate. The x-axis is the common normal to all layers as shown in **Figure 1**. The thickness of the considered multilayered structure and the ith layer is denoted by h and h^i, respectively. The considered multilayered structure which occupies the region $R = \{(x, y, z) : 0 < x < h, 0 < y < b, 0 < z < a\}$ has been placed in a primary magnetic field H_0 acting in the direction of the y-axis.

According to the three-temperature theory, the governing equations of nonlinear generalized magneto-thermoelasticity in an initially stressed multilayered functionally graded anisotropic (ISMFGA) structure for the ith layer can be written in the following form:

$$\sigma_{ab,b} + \tau_{ab,b} - \Gamma_{ab} = \rho^i (x+1)^m \ddot{u}_a^i \tag{1}$$

$$\sigma_{ab} = (\chi+1)^m \left[C^i_{abfg} u^i{}_{f,g} - \beta^i_{ab}\left(T^i - T_0 + \tau_1 \dot{T}^i\right)\right] \tag{2}$$

$$\tau_{ab} = \mu^i (x+1)^m \left(\tilde{h}_a H_b + \tilde{h}_b H_a - \delta_{ba}\left(\tilde{h}_f H_f\right)\right) \tag{3}$$

$$\Gamma_{ab} = P^i (x+1)^m \left(\frac{\partial u^i_a}{\partial x_b} - \frac{\partial u^i_b}{\partial x_a}\right) \tag{4}$$

According to Fahmy [10], the *2D-3 T radiative heat conduction equations* can be expressed as follows:

$$\nabla\left[\left(\delta_{1j}\mathbb{K}^{i*}_\alpha + \delta_{2j}\mathbb{K}^i_\alpha\right)\nabla T^i_\alpha(r,\tau)\right] - \overline{\mathbb{W}}(r,\tau) = c^i_\alpha \rho^i \delta_1 \delta_{1j} \frac{\partial T^i_\alpha(r,\tau)}{\partial \tau} \tag{5}$$

where

$$\overline{\mathbb{W}}(r,\tau) = \begin{cases} \rho^i \mathbb{W}_{eI}\left(T^i_e - T^i_I\right) + \rho^i \mathbb{W}_{er}\left(T^i_e - T^i_p\right) + \overline{\overline{\mathbb{W}}}, \alpha = e, & \delta_1 = 1 \\ -\rho^i \mathbb{W}_{eI}\left(T^i_e - T^i_p\right) + \overline{\overline{\mathbb{W}}}, & \alpha = I, \delta_1 = 1 \\ -\rho^i \mathbb{W}_{er}\left(T^i_e - T^i_p\right) + \overline{\overline{\mathbb{W}}}, & \alpha = p, \delta_1 = T^3_p \end{cases} \tag{6}$$

in which

$$\overline{\overline{\mathbb{W}}}(r,\tau) = -\delta_{2j}\mathbb{K}^i_\alpha \dot{T}_{\alpha,ab} + \beta_{ab} T_{\alpha 0}\left[\left(\tau_0 + \delta_{2j}\right)\ddot{u}_{a,b}\right] + \rho^i c^i_\alpha\left[\left(\tau_0 + \delta_{1j}\tau_2 + \delta_{2j}\right)\ddot{T}_\alpha\right] - Q(x,\tau) \tag{7}$$

and

$$\begin{aligned} &\mathbb{W}_{eI} = \rho^i \mathbb{A}_{eI} T_e^{-2/3}, \mathbb{W}_{er} = \rho^i \mathbb{A}_{er} T_e^{-1/2}, \mathbb{K}_\alpha = \mathbb{A}_\alpha T_\alpha^{5/2}, \alpha = e, I, \\ &\mathbb{K}_p = \mathbb{A}_p T_p^{3+\mathbb{B}} \end{aligned} \tag{8}$$

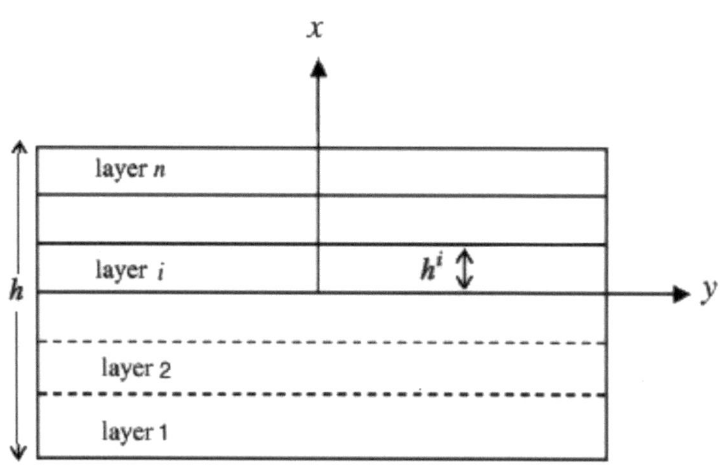

Figure 1.
Geometry of the FGA structure.

The total energy of unit mass can be described by

$$P = P_e + P_I + P_p,\ P_e = c_{\alpha e} T_e^i,\ P_I = c_{\alpha I} T_l^i,\ P_p = \frac{1}{4} c_{\alpha p} T_p^{4i} \tag{9}$$

where σ_{ab}, τ_{ab}, and u_k^i are the mechanical stress tensor, Maxwell's electromagnetic stress tensor, and displacement vector, respectively; $T_{\alpha 0}^i$ is the reference temperature; T_α^i is the temperature; C_{abfg}^i and β_{ab}^i are, respectively, the constant elastic moduli and stress-temperature coefficients of the anisotropic medium; μ^i, $\tilde{h}$, P^i, ρ^i, and c_α^i are the magnetic permeability, perturbed magnetic field, initial stress in the ith layer, density, and specific heat capacity, respectively; τ is the time; τ_0, τ_1, and τ_2 are the relaxation times; $i = 1, 2, \ldots, n-1$ represents the parameters in a multilayered structure; and m is a dimensionless constant. Also, we considered in the current study that $\tau_{ab,b} = \mu_0^i \epsilon_{abf} J_b H_f$ is the a-component of the Lorentz force and $J(\tau) = \frac{J_0 \tau}{\tau_3^2} e^{\frac{\tau}{\tau_3}}$ is the temporal profile of a non-Gaussian laser pulse, J_0 is the total energy intensity, and $Q(x,\tau) = \frac{1-R}{x_0} e^{\left(\frac{x_a}{x_0}\right) J(\tau)}, a = 1, 2, 3$ is the heat source intensity.

According to Fahmy [57], we notice that there are two special cases of the Green and Naghdi theory of type III; when $\mathbb{K}_\alpha^i \to 0$, the equations of GN III theory are reduced to the GN theory type II, and when $\mathbb{K}_\alpha^{i*} \to 0$, the equations of the GN III theory are reduced to the GN theory type I.

3. Continuity and initial and boundary conditions

The continuity conditions along interfaces for the temperature, heat flux, displacement, and traction can be expressed as follows:

$$T_\alpha^i(x,z,\tau)\big|_{x=h^i} = T_\alpha^{(i+1)}(x,z_t\tau)\big|_{x=h^i} \tag{10}$$

$$q^i(x,z,\tau)\big|_{x=h^i} = q^{(i+1)}(x,z,\tau)\big|_{x=h^i} \tag{11}$$

$$u_f^i(x,z,\tau)\Big|_{x=h^i} = u_f^{(i+1)}(x,z,\tau)_{x=h^i} \tag{12}$$

$$\bar{t}_a^i(x,z,\tau)\Big|_{x=h^i} = \bar{t}_a^{(i+1)}(x,z,\tau)_{x=h^i} \tag{13}$$

where n is the total number of layers, $\bar{t}_a$ are the tractions, which are defined by $\bar{t}_a = \sigma_{ab} n_b$, and $i = 1, 2, \ldots, n-1$.

The remaining initial and boundary conditions for the current study are

$$u_f^i(x,z,0) = \dot{u}_f^i(x,z,0) = 0 \quad \text{for } (x,z) \in R \cup C \tag{14}$$

$$u_f^i(x,z,\tau) = \Psi_f(x,z,\tau) \ \text{ for } (x,z) \in C_3 \tag{15}$$

$$\bar{t}_a^i(x,z,\tau) = \Phi_f(x,z,\tau) \ \text{ for } (x,z) \in C_4, \tau > 0, \tag{16}$$

$$T_\alpha^i(x,z,0) = T_\alpha^i(x,z,0) = 0 \ \text{ for } (x,z) \in R \cup C \tag{17}$$

$$T_\alpha^i(x,y,\tau) = \bar{f}(x,y,\tau) \ \text{ for } (x,y) \in C_1, \quad \tau > 0 \tag{18}$$

$$q^i(x,z,\tau) = \bar{h}(x,z,\tau) \ \text{ for } (x,z) \in C_2, \tau > 0 \tag{19}$$

where Ψ_f, Φ_f, f, and $\overline{h}$ are suitably prescribed functions and $C = C_1 \cup C_2 = C_3 \cup C_4$, $C_1 \cap C_2 = C_3 \cap C_4 = \emptyset$.

4. BEM numerical implementation

Making use of Eqs. (2)–(4), we can write (1) as follows:

$$L_{gb} u^i_f = \rho^i \ddot{u}^i_a - \left(D_a T^i_\alpha - P^i \left(\frac{\partial u^i_b}{\partial x_a} - \frac{\partial u^i_a}{\partial x_b} \right) \right) = f_{gb} \tag{20}$$

where the inertia term $\rho \ddot{u}_a$, the temperature gradient $D_a T$, and the initial stress term are treated as the body forces.

The field equations may be expressed in the operator form as follows:

$$L_{gb} u^i_f = f_{gb}, \tag{21}$$

$$L_{ab} T^i_\alpha = f_{ab} \tag{22}$$

where the operators L_{gb}, f_{gb}, L_{ab}, and f_{ab} are as follows:

$$L_{gb} = D_{abf} \frac{\partial}{\partial x_b} + D_{af} + \Lambda D_{a1f}, \qquad L_{ab} = \left(\delta_{2j} \mathbb{K}^{i*}_\alpha \right) \nabla \tag{23}$$

$$f_{gb} = \rho^i \ddot{u}^i_a - \left(D_a T^i_\alpha - P^i \left(\frac{\partial u^i_b}{\partial x_a} - \frac{\partial u^i_a}{\partial x_b} \right) \right) \tag{24}$$

$$f_{ab} = \nabla \left(\delta_{1j} \mathbb{K}^i_\alpha \right) \nabla + \rho^i c^i_\alpha \delta_1 \delta_{1j} (x+1)^m \dot{T}^i_\alpha + \overline{\mathbb{W}}(r, \tau) \tag{25}$$

where

$$D_{abf} = C_{abfg} \varepsilon, \varepsilon = \frac{\partial}{\partial x_g}, D_{af} = \mu H_0^2 \left(\frac{\partial}{\partial x_a} + \delta_{a1} \Lambda \right) \frac{\partial}{\partial x_f},$$

$$D_a = -\beta^i_{ab} \left(\frac{\partial}{\partial x_b} + \delta_{b1} \Lambda + \tau_1 \left(\frac{\partial}{\partial x_b} + \Lambda \right) \frac{\partial}{\partial \tau} \right), \qquad \Lambda = \frac{m}{x+1}.$$

The differential equation (21) can be solved using the weighted residual method (WRM) to obtain the following integral equation:

$$\int_R \left(L_{gb} u^i_f - f_{gb} \right) u^{i*}_{da} dR = 0 \tag{26}$$

Now, the fundamental solution u^{i*}_{df} and traction vectors t^{i*}_{da} and t^i_a can be written as follows:

$$L_{gb} u^{i*}_{af} = -\delta_{ad} \delta(x, \xi) \tag{27}$$

$$t^{i*}_{da} = C_{abfg} u^{i*}_{df,g} n_b \tag{28}$$

$$t^i_a = \frac{\overline{t}^i_a}{(x+1)^m} = \left(C_{abfg} u^i_{f,g} - \beta^i_{ab} \left(T^i_\alpha + \tau_1 \dot{T}^i_\alpha \right) \right) n_b \tag{29}$$

Using integration by parts and sifting property of the Dirac distribution for (26), then using Eqs. (27) and (29), we can write the following elastic integral representation formula:

$$u_d^i(\xi) = \int_C \left(u_{da}^{i*} t_a^i - t_{da}^{i*} u_a^i + u_{da}^{i*} \beta_{ab}^i T_\alpha^i n_b\right) dC - \int_R f_{gb} u_{da}^{i*} dR \tag{30}$$

The fundamental solution T^{i^*}can be defined as

$$L_{ab} T^{i^*} = -\delta(x, \xi) \tag{31}$$

By using WRM and integration by parts, we can write (23) as follows:

$$\int_R \left(L_{ab} T_\alpha^i T_\alpha^{i^*} - L_{ab} T_\alpha^{i^*} T_\alpha^i\right) dR = \int_C \left(q^{i^*} T_\alpha^i - q^i T_\alpha^{i^*}\right) dC \tag{32}$$

where

$$q^i = -\mathbb{K}_\alpha^i T_{\alpha.b}^i n_a \tag{33}$$

$$q^{i*} = -\mathbb{K}_\alpha^i T_{\alpha,b}^{i^*} n_a \tag{34}$$

By the use of sifting property, we obtain from (32) the thermal integral representation formula:

$$T_\alpha^i(\xi) = \int_C \left(q^{i^*} T_\alpha^i - q^i T_\alpha^{i^*}\right) dC - \int_R f_{ab} T_\alpha^{i^*} dR \tag{35}$$

By combining (30) and (35), we have

$$\begin{bmatrix} u_d^i(\xi) \\ T_\alpha^i(\xi) \end{bmatrix} = \int_C \left\{ -\begin{bmatrix} t_{da}^{i*} & -u_{aa}^{i*} \beta_{ab} n_b \\ 0 & -q^{i*} \end{bmatrix} \begin{bmatrix} u_a^i \\ T_\alpha^i \end{bmatrix} + \begin{bmatrix} u_{da}^{i*} & 0 \\ 0 & -T_\alpha^{i*} \end{bmatrix} \begin{bmatrix} \tau_a^i \\ q^i \end{bmatrix} \right\} dC - \int_R \begin{bmatrix} u_{da}^{i*} & 0 \\ 0 & -T_\alpha^{i*} \end{bmatrix} \begin{bmatrix} f_{gb} \\ -f_{ab} \end{bmatrix} dR \tag{36}$$

The generalized thermoelastic vectors can be expressed in contracted notation form as follows:

$$U_A^i = \begin{cases} u_a^i & a = A = 1, 2, 3 \\ T_\alpha^i & A = 4 \end{cases} \tag{37}$$

$$\mathrm{T}_{\alpha A}^i = \begin{cases} t_a^i & a = A = 1, 2, 3 \\ q^i & A = 4 \end{cases} \tag{38}$$

$$U_{DA}^{i*} = \begin{cases} u_{da}^{i*} & d = D = 1, 2, 3; a = A = 1, 2, 3 \\ 0 & d = D = 1, 2, 3; A = 4 \\ 0 & D = 4; a = A = 1, 2, 3 \\ -T_\alpha^{i*} & D = 4; A = 4 \end{cases} \tag{39}$$

$$\tilde{T}_{\alpha DA}^{i^*} = \begin{cases} t_{aa}^{i*} & d = D = 1, 2, 3; a = A = 1, 2, 3 \\ -\tilde{u}_d^{i*} & d = D = 1, 2, 3; A = 4 \\ 0 & D = 4; a = A = 1, 2, 3 \\ -q^{i*} & D = 4; A = 4 \end{cases} \tag{40}$$

$$\tilde{u}_d^{i*} = u_{da}^{i*}\beta_{af}^{i} n_f \tag{41}$$

Using the previous vectors, we can write (36) as

$$U_D^i(\xi) = \int_C \left(U_{DA}^{i*}\mathrm{T}_{\alpha A}^{i} - \tilde{T}_{\alpha DA}^{i} U_A^i\right) dC - \int_R U_{DA}^{i*} S_A dR \tag{42}$$

The vector S_A can be split as follows

$$S_A = S_A^0 + S_A^T + S_A^{\mathrm{u}} + S_A^{\dot{T}} + S_A^{\ddot{T}} + S_A^{\ddot{\mathrm{u}}} \tag{43}$$

where

$$S_A^0 = \begin{cases} 0 & A = 1,2,3 \\ \dfrac{1-R}{x_0} e^{\left(-\frac{x_a}{x_0}\right)J(\tau)} & A = 4 \end{cases} \tag{44}$$

$$S_A^T = \omega_{AF} U_F^i \text{with} \omega_{AF} = \begin{cases} -D_a & A = 1,2,3; F = 4 \\ \nabla\left(\delta_{2j}\mathbb{K}_\alpha^{i^*}\right)\nabla & \text{otherwise} \end{cases} \tag{45}$$

$$S_A^u = \psi U_F^i \text{with} \psi = \begin{cases} P^i\left(\dfrac{\partial}{\partial xb} - \dfrac{\partial}{\partial x_a}\right) & A = 1,2,3; F = 1,2,3, \\ 0 & A = 4; F = 4 \end{cases} \tag{46}$$

$$S_A^{\dot{T}} = \Gamma_{AF}\dot{U}_F^i \quad \text{with} \quad \Gamma_{AF} = \begin{cases} -\beta_{ab}^i \tau_1\left(\dfrac{\partial}{\partial x_b} + \Lambda\right)\dfrac{\partial}{\partial \tau} & A = 4; F = 4 \\ \rho^i c_\alpha^i \delta_1 \delta_{1j} & \text{otherwise} \end{cases} \tag{47}$$

$$S_A^{\ddot{T}} = \delta_{AF}\ddot{U}_F^i \quad \text{with} \quad \delta_{AF} = \begin{cases} 0 & A = 4; F = 4 \\ \rho^i c_\alpha^i\left[\left(\tau_0 + \delta_{1j}\tau_2 + \delta_{2j}\right)\right] & \text{otherwise} \end{cases} \tag{48}$$

$$S_A^{\ddot{\mathrm{u}}} = \tilde{\sigma}\ddot{U}_F^i \text{ with } \tilde{\sigma} = \begin{cases} \rho^i & A = 1,2,3, F = 1,2,3, \\ \beta_{ab}^i T_{\alpha 0}^i\left(\tau_0 + \delta_{2i}\right) & A = 4; F = 4 \end{cases} \tag{49}$$

The thermoelastic representation formula (36) can also be written in matrix form as follows:

$$\begin{aligned}
[S_A] = &- \begin{bmatrix} 0 \\ -\dfrac{1-R}{x_0} e\left(-\dfrac{x_a}{x_0}\right)J(\tau) \end{bmatrix} + \left\{\begin{bmatrix} -D_a T_\alpha^i \\ \nabla\left(\delta_{2j}\mathbb{K}_\alpha^{i^*}\right)\nabla T_\alpha^i \end{bmatrix}\right\} + \begin{bmatrix} P^i\left(u_{b,a}^{i_0} - u_{a,b}^i\right) \\ 0 \end{bmatrix} \\
&+ \begin{bmatrix} -\beta_{ab}^i \tau_1\left(\dfrac{\partial}{\partial x_b} + \Lambda\right)\dot{T}_\alpha^i \\ \rho^i c_\alpha^i \delta_1 \delta_{1j}\dot{T}_\alpha^i \end{bmatrix} + \rho^i c_\alpha^i\left[\left(\tau_0 + \delta_{1j}\tau_2 + \delta_{2j}\right)\right]\begin{bmatrix} 0 \\ \ddot{T}_\alpha^i \end{bmatrix} \\
&+ \begin{bmatrix} \rho^i \ddot{u}_a^i \\ \beta_{ab}^i T_{\alpha 0}^i\left(\tau_0 + \delta_{2j}\right)\ddot{u}_{f,g}^i \end{bmatrix}
\end{aligned} \tag{50}$$

To transform the domain integral in (42) to the boundary, we approximate the source vector S_A by a series of given tensor functions f_{AE}^q and unknown coefficients α_E^q as follows:

$$S_A \approx \sum_{q=1}^{E} f^q_{AE} \alpha^q_E \tag{51}$$

Thus, the thermoelastic representation formula (42) can be written in the following form:

$$U_D(\xi) = \int_C \left(U^{i^*}_{DA} T^i_{\alpha A} - \tilde{T}^{i^*}_{\alpha DA} U^i_A \right) dC - \sum_{q=1}^{N} \int_R U^{i^*}_{DA} f^q_{AE} dR \, \alpha^q_E \tag{52}$$

By implementing the WRM to the following equations.

$$L_{gb} u^{iq}_{fe} = f^q_{ae} \tag{53}$$

$$L_{ab} T^{iq}_{\alpha} = f^q_{pj} \tag{54}$$

Then, the elastic and thermal representation formulae are given as follows [46]:

$$u^{iq}_{de}(\xi) = \int_C (u^{i*}_{da} t^{iq}_{ae} - t^{i*}_{da} u^{iq}_{ae}) dC - \int_R u^{i*}_{da} f^q_{ae} dR \tag{55}$$

$$T^{iq}_{\alpha}(\xi) = \int_C (q^{i*} T^{iq}_{\alpha} - q^{iq} T^{i*}_{\alpha}) \, dC - \int_R f^q T^{i*}_{\alpha} \, dR \tag{56}$$

The representation formulae (55) and (56) can be combined into the following single equation:

$$U^{iq}_{DE}(\xi) = \int_C \left(U^{i*}_{DA} T^{iq}_{\alpha AE} - T^{i*}_{\alpha DA} U^{iq}_{AE} \right) dC - \int_R U^{i*}_{DA} f^{iq}_{AE} dR \tag{57}$$

With the substitution of (57) into (52), the dual reciprocity representation formula of coupled thermoelasticity can be expressed as follows:

$$\begin{aligned} U^i_D(\xi) = {} & \int_C \left(U^{i*}_{DA} T^i_{\alpha A} - \breve{T}^{i*}_{\alpha DA} U^i_A \right) dC \\ & + \sum_{q=1}^{E} \left(U^{iq}_{DE}(\xi) + \int_C \left(T^{i*}_{\alpha DA} U^{iq}_{AE} - U^{i*}_{DA} T^{iq}_{\alpha AE} \right) dC \right) \alpha^q_E \end{aligned} \tag{58}$$

To calculate interior stresses, (58) is differentiated with respect to ξ_l as follows:

$$\begin{aligned} \frac{\partial U^i_D(\xi)}{\partial \xi_l} = {} & -\int_C \left(U^{i*}_{DA,l} T^i_{\alpha A,l} - \breve{T}^{i^*}_{\alpha DA,l} U^i_A \right) dC \\ & + \sum_{q=1}^{E} \left(\frac{\partial U^{iq}_{DE}(\xi)}{\partial \xi_l} - \int_C \left(T^{i*}_{\alpha DA,l} U^{iq}_{\alpha AE} - U^{i*}_{DA,l} T^{iq}_{\alpha AE} \right) dC \right) \alpha^q_E \end{aligned} \tag{59}$$

According to the dual reciprocity boundary integral equation procedure of Fahmy [44], we can write (58) in the following system of equations:

$$\breve{\zeta}U - \eta T_\alpha = \left(\zeta\breve{U} - \eta\breve{\wp}\right)\alpha \tag{60}$$

The generalized displacements and velocities are approximated in terms of a series of known tensor functions f^q_{FD} and unknown coefficients γ^q_D and $\tilde{\gamma}^q_D$:

$$U^i_F \approx \sum_{q=1}^{N} f^q_{FD}(x)\gamma^q_D \tag{61}$$

where

$$f^q_{FD} = \begin{cases} f^q_{fd} & f = F = 1,2,3; d = D = 1,2,3 \\ f^q & F = 4; D = 4 \\ 0 & \text{otherwise} \end{cases} \tag{62}$$

The gradients of the generalized displacements and velocities can also be approximated in terms of the derivatives of tensor functions as follows:

$$U^i_{F,g} \approx \sum_{q=1}^{N} f^q_{FD,g}(x)\gamma^q_K \tag{63}$$

These approximations are substituted into Eq. (45) to obtain

$$S^T_A = \sum_{q=1}^{N} S_{AF} f^q_{FD,g}\gamma^q_D \tag{64}$$

By implementing the point collocation procedure introduced by Gaul et al. [43] to Eqs. (51) and (61), we have

$$\breve{S} = J\overline{\alpha}, \quad U^i = J'\gamma, \tag{65}$$

Similarly, the implementation of the point collocation procedure to Eqs. (64), (46), (47), (48), and (49) leads to the following equations:

$$\breve{S}^{T^i_\alpha} = \mathcal{B}^T\gamma \tag{66}$$

$$S^u_A = \overline{\psi}U^i \tag{67}$$

$$\breve{S}^{\dot{T}^i_\alpha} = \overline{\Gamma}_{AF}\dot{U}^i \tag{68}$$

$$\breve{S}^{\ddot{T}^i_\alpha} = \overline{\delta}_{AF}\ddot{U}^i \tag{69}$$

$$\breve{S}^{\ddot{u}} = \overline{\tilde{o}}\ddot{U}^i \tag{70}$$

where $\overline{\psi}$, $\overline{\Gamma}_{AF}$, $\overline{\delta}_{AF}$, and $\overline{\tilde{o}}$ are assembled using the submatrices $[\psi]$, $[\Gamma_{AF}]$, $[\delta_{AF}]$, and $[\tilde{o}]$, respectively.

Solving the system (65) for α and γ yields

$$\overline{\alpha} = J^{-1}\breve{S}, \quad \gamma = J'^{-1}U^i \tag{71}$$

Now, the coefficients α can be expressed in terms of nodal values of the unknown displacements U^i, velocities $\dot{U}^i$, and accelerations $\ddot{U}^i$ as follows:

$$\overline{\alpha} = J^{-1}\left(\breve{S}^0 + \left(\mathcal{B}^T J'^{-1} + \overline{\psi}\right]U^i + \overline{\Gamma}_{AF}\dot{U}^i + \left(\overline{o} + \overline{\delta}_{AF}\right)\ddot{U}^i\right) \tag{72}$$

An implicit-implicit staggered algorithm for the integration of the governing equations was developed and implemented for use with the DRBEM for solving the governing equations which may now be written in a more convenient form after substitution of Eq. (72) into Eq. (60) as follows:

$$\widehat{M}\ddot{U}^i + \widehat{\Gamma}\dot{U}^i + \widehat{K}U^i = \widehat{\mathbb{Q}}^i \tag{73}$$

$$\widehat{\mathrm{X}}\ddot{T}^i_\alpha + \widehat{\mathrm{A}}\dot{T}^i_\alpha + \widehat{\mathrm{B}}T^i_\alpha = \widehat{\mathbb{Z}}\ddot{U}^i + \widehat{\mathbb{R}} \tag{74}$$

where $V = \left(\eta\breve{\wp} - \zeta\breve{U}\right)J^{-1}$, $\widehat{M} = V\left(\overline{o} + \overline{\delta}_{AF}\right)$, $\widehat{\Gamma} = V\overline{\Gamma}_{AF}$,
$\widehat{K} = -\breve{\zeta} + V\left(\mathcal{B}^T J\prime^{-1} + \overline{\psi}\right)$, $\widehat{\mathbb{Q}}^i = -\eta\breve{T} + V\breve{S}^0$, $\widehat{\mathrm{X}} = -\rho^i c^i (x+1)^m$,
$\widehat{\mathrm{A}} = k^i_{ab}\frac{\partial}{\partial x_a}\frac{\partial}{\partial x_b}$, $\widehat{\mathrm{B}} = k^{i^*}_{ab}\frac{\partial}{\partial x_a}\frac{\partial}{\partial x_b}$, $\widehat{\mathbb{Z}} = \beta^i_{ab}T_0$, $\widehat{\mathbb{R}} = -\frac{1-R}{x_0}e^{\left(\frac{x_a}{x_0}\right)J(\tau)}$
where V, $\widehat{M}$, $\widehat{\Gamma}$, $\widehat{K}$, $\widehat{\mathrm{A}}$, and $\widehat{\mathrm{B}}$ represent the volume, mass, damping, stiffness, capacity, and conductivity matrices, respectively, and $\ddot{U}^i$, $\dot{U}^i$, U^i, T^i, and $\widehat{\mathbb{Q}}^i$ represent the acceleration, velocity, displacement, temperature, and external force vectors, respectively.

In many applications, the coupling term $\widehat{\mathbb{Z}}\ddot{U}^i_{n+1}$ that appears in the heat conduction equation and which is induced by the effect of the strain rate is negligible.

Hence, Eqs. (73) and (74) lead to the following coupled system of differential-algebraic equations (DAEs):

$$\widehat{M}\ddot{U}^i_{n+1} + \widehat{\Gamma}\dot{U}^i_{n+1} + \widehat{K}U^i_{n+1} = \widehat{\mathbb{Q}}^{ip}_{n+1} \tag{75}$$

$$\widehat{\mathrm{X}}\ddot{T}^i_{\alpha(n+1)} + \widehat{\mathrm{A}}\dot{T}^i_{\alpha(n+1)} + \widehat{\mathrm{B}}T^i_{\alpha(n+1)} = \widehat{\mathbb{Z}}\ddot{U}^i_{n+1} + \widehat{\mathbb{R}} \tag{76}$$

where $\widehat{\mathbb{Q}}^{ip}_{n+1} = \eta T^{ip}_{\alpha(n+1)} + V\breve{S}^0$ and $T^{ip}_{\alpha(n+1)}$ is the predicted temperature.
Integrating Eq. (73) with the use of trapezoidal rule and Eq. (75), we obtain

$$\begin{aligned} U^i_{n+1} &= \dot{U}^i_n + \frac{\Delta\tau}{2}\left(\ddot{U}^i_{n+1} + \ddot{U}^i_n\right) \\ &= \dot{U}^i_n + \frac{\Delta\tau}{2}\left[\ddot{U}^i_n + \widehat{M}^{-1}\left(\widehat{\mathbb{Q}}^{ip}_{n+1} - \widehat{\Gamma}\dot{U}^i_{n+1} - \widehat{K}U^i_{n+1}\right)\right] \end{aligned} \tag{77}$$

$$\begin{aligned} U^i_{n+1} &= U^i_n + \frac{\Delta\tau}{2}\left(\dot{U}^i_{n+1} + \dot{U}^i_n\right) \\ &= U^i_n + \Delta\tau\dot{U}^i_n + \frac{\Delta\tau^2}{4}\left[\ddot{U}^i_n + \widehat{M}^{-1}\left(\tilde{\mathbb{Q}}^{ip}_{n+1} - \widehat{\Gamma}\dot{U}^i_{n+1} - \widehat{K}U^i_{n+1}\right)\right] \end{aligned} \tag{78}$$

From Eq. (77) we have

$$\dot{U}^{i}_{n+1} = \overline{\Upsilon}^{-1}\left[\dot{U}^{i}_{n} + \frac{\Delta\tau}{2}\left[\ddot{U}^{i}_{n} + \widehat{M}^{-1}\left(\widehat{\mathbb{Q}}^{ip}_{n+1} - \widehat{K}U^{i}_{n+1}\right)\right]\right] \tag{79}$$

where $\overline{\Upsilon} = \left(I + \frac{\Delta\tau}{2}\widehat{M}^{-1}\widehat{\Gamma}\right)$.
Substituting from Eq. (79) into Eq. (78), we derive

$$\begin{aligned} U^{i}_{n+1} = U^{i}_{n} + \Delta\tau\dot{U}^{i}_{n} + \frac{\Delta\tau^2}{4}\Bigg[\ddot{U}^{i}_{n} + \widehat{M}^{-1}\Bigg(\tilde{\mathbb{Q}}^{ip}_{n+1} - \widehat{\Gamma}\,\overline{\Upsilon}^{-1} \\ \left[\dot{U}^{i}_{n} + \frac{\Delta\tau}{2}\right]\left[\ddot{U}^{i}_{n} + \widehat{M}^{-1}\left(\widehat{\mathbb{Q}}^{ip}_{n+1} - \widehat{K}U^{i}_{n+1}\right)\right] - \widehat{K}U^{i}_{n+1}\Bigg)\Bigg] \end{aligned} \tag{80}$$

Substituting $\dot{U}^{i}_{n+1}$ from Eq. (79) into Eq. (75), we obtain

$$\ddot{U}^{i}_{n+1} = \widehat{M}^{-1}\left[\widehat{\mathbb{Q}}^{ip}_{n+1} - \widehat{\Gamma}\left[\overline{\Upsilon}^{-1}\left[\dot{U}^{i}_{n} + \frac{\Delta\tau}{2}\left[\ddot{U}^{i}_{n} + \widehat{M}^{-1}\left(\widehat{\mathbb{Q}}^{ip}_{n+1} - \widehat{K}U^{i}_{n+1}\right)\right]\right]\right] - \widehat{K}U^{i}_{n+1}\right] \tag{81}$$

Integrating the heat Eq. (74) using the trapezoidal rule and Eq. (76), we get

$$\begin{aligned} \dot{T}^{i}_{\alpha(n+1)} &= \dot{T}^{i}_{n} + \frac{\Delta\tau}{2}\left(\ddot{T}^{i}_{\alpha(n+1)} + \ddot{T}^{i}_{\alpha n}\right) \\ &= \dot{T}^{i}_{\alpha n} + \frac{\Delta\tau}{2}\left(\widehat{\mathrm{X}}^{-1}\left[\widehat{\mathbb{Z}}\ddot{U}^{i}_{n+1} + \widehat{\mathbb{R}} - \widehat{\mathrm{A}}\dot{T}^{i\mathrm{A}}_{\alpha(n+1)} - \widehat{\mathrm{B}}T^{i}_{\alpha(n+1)}\right] + \ddot{T}^{i}_{\alpha n}\right) \end{aligned} \tag{82}$$

$$\begin{aligned} T^{i}_{\alpha(n+1)} &= T^{i}_{\alpha n} + \frac{\Delta\tau}{2}\left(\dot{T}^{i}_{\alpha(n+1)} + \dot{T}^{i}_{\alpha n}\right) \\ &= T^{i}_{\alpha n} + \Delta\tau\dot{T}^{i}_{\alpha n} \\ &\quad + \frac{\Delta\tau^2}{4}\left(\ddot{T}^{i}_{\alpha n} + \widehat{\mathrm{X}}^{-1}\left[\widehat{\mathbb{Z}}\ddot{U}^{i}_{n+1} + \widehat{\mathbb{R}} - \widehat{\mathrm{A}}\dot{T}^{i}_{\alpha(n+1)} - \widehat{\mathrm{B}}T^{i}_{\alpha(n+1)}\right]\right) \end{aligned} \tag{83}$$

From Eq. (82) we get

$$\dot{T}^{i}_{\alpha(n+1)} = \Upsilon^{-1}\left[\dot{T}^{i}_{\alpha n} + \frac{\Delta\tau}{2}\left(\widehat{\mathrm{X}}^{-1}\left[\widehat{\mathbb{Z}}\ddot{U}^{i}_{n+1} + \widehat{\mathbb{R}} - \widehat{\mathrm{B}}T^{i}_{\alpha(n+1)}\right] + \ddot{T}^{i}_{\alpha n}\right)\right] \tag{84}$$

where $\Upsilon = \left(I + \frac{1}{2}\widehat{\mathrm{A}}\,\Delta\tau\,\widehat{\mathrm{X}}^{-1}\right)$
Substituting from Eq. (84) into Eq. (83), we have

$$\begin{aligned} T^{i}_{\alpha(n+1)} = T^{i}_{\alpha n} + \Delta\tau\dot{T}^{i}_{\alpha n} + \frac{\Delta\tau^2}{4}\Bigg(\ddot{T}^{i}_{\alpha n} + \widehat{\mathrm{X}}^{-1}\Bigg[\widehat{\mathbb{Z}}\ddot{U}^{i}_{n+1} + \widehat{\mathbb{R}} - \widehat{\mathrm{A}}\Bigg(\mathrm{Y}^{-1}\left[\dot{T}^{i}_{\alpha n}\right] \\ + \frac{\Delta\tau}{2}\left(\widehat{\mathrm{X}}^{-1}\left[\widehat{\mathbb{Z}}\ddot{U}^{i}_{n+1} + \widehat{\mathbb{R}} - \widehat{\mathrm{B}}T^{i}_{\alpha(n+1)}\right] + \ddot{T}^{i}_{\alpha n}\right)\Bigg) - \tilde{\mathrm{B}}T^{i}_{\alpha(n+1)}\Bigg]\Bigg) \end{aligned} \tag{85}$$

Substituting $\dot{T}^{i}_{n+1}$ from Eq. (84) into Eq. (76), we obtain

$$\ddot{T}_{\alpha(n+1)}i^{\mathrm{A}} = \widehat{\mathrm{X}}^{-1}\left[\widehat{\mathbb{Z}}\,\ddot{U}^{i}_{n+1} + \widehat{\mathbb{R}} - \widehat{\mathrm{A}}\left(\Upsilon^{-1}\left[\dot{T}^{i}_{\alpha n} + \frac{\Delta\tau}{2}\left(\widehat{\mathrm{X}}^{-1}\left[\widehat{\mathbb{Z}}\,\ddot{U}^{i}_{n+1} + \widehat{\mathbb{R}} - \widehat{\mathrm{B}}\,T^{i}_{\alpha(n+1)}\right] + \ddot{T}^{i}_{\alpha(n+1)}\right)\right]\right) - *\,\widehat{\mathrm{B}}\,T^{i}_{\alpha(n+1)}\right] \tag{86}$$

Now, a displacement-predicted staggered procedure for the solution of (80) and (85) is as follows:

The first step is to predict the propagation of the displacement wave field: $U^{ip}_{n+1} = U^{i}_{n}$. The second step is to substitute $\dot{U}^{i}_{n+1}$ and $\ddot{U}^{i}_{n+1}$ from Eqs. (77) and (75), respectively, in Eq. (85) and solve the resulting equation for the three-temperature wave fields. The third step is to correct the displacement wave propagation using the computed three-temperature fields for Eq. (80). The fourth step is to compute $\dot{U}^{i}_{n+1}$, $\ddot{U}^{i}_{n+1}$, $\dot{T}^{i}_{\alpha(n+1)}$, and $\ddot{T}^{i}_{\alpha(n+1)}$ from Eqs. (79), (81), (82), and (86), respectively.

5. Numerical results and discussion

In order to show the numerical results of this study, we consider a monoclinic graphite-epoxy as an anisotropic thermoelastic material which has the following physical constants [57]:

The elasticity tensor is expressed as

$$C_{pjkl} = \begin{bmatrix} 430.1 & 130.4 & 18.2 & 0 & 0 & 201.3 \\ 130.4 & 116.7 & 21.0 & 0 & 0 & 70.1 \\ 18.2 & 21.0 & 73.6 & 0 & 0 & 2.4 \\ 0 & 0 & 0 & 19.8 & -8.0 & 0 \\ 0 & 0 & 0 & -8.0 & 29.1 & 0 \\ 201.3 & 70.1 & 2.4 & 0 & 0 & 147.3 \end{bmatrix} \mathrm{GPa} \tag{87}$$

The mechanical temperature coefficient is

$$\beta_{pj} = \begin{bmatrix} 1.01 & 2.00 & 0 \\ 2.00 & 1.48 & 0 \\ 0 & 0 & 7.52 \end{bmatrix} \cdot 10^{6}\,\frac{\mathrm{N}}{\mathrm{Km}^2} \tag{88}$$

The thermal conductivity tensor is

$$k_{pj} = \begin{bmatrix} 5.2 & 0 & 0 \\ 0 & 7.6 & 0 \\ 0 & 0 & 38.3 \end{bmatrix} \mathrm{W/Km} \tag{89}$$

Mass density $\rho = 7820\,\mathrm{kg/m^3}$ and heat capacity $c = 461\,\mathrm{J/kg\,K}$.

The proposed technique that has been utilized in the present chapter can be applicable to a wide range of laser wave propagations in three-temperature nonlinear generalized thermoelastic problems of FGA structures. The main aim of this paper was to assess the impact of three temperatures on the acoustic displacement waves; the numerical outcomes are completed and delineated graphically for electron, ion, phonon, and total temperatures.

Figure 2 shows the three temperatures T_e, T_i, and T_p and total temperature $T(T = T_e + T_i + T_p)$ wave propagation along the x-axis. It was shown from this figure that the three temperatures are different and they may have great effects on the connected fields.

Figures 3 and **4** show the displacement u_1 and u_2 acoustic waves propagation along x-axis for the three temperatures T_e, T_i, T_p and total temperature T. It was noticed from **Figures 3** and **4** that the three temperatures and total temperature have great effects on the acoustic displacement waves.

In order to evaluate the influence of the functionally graded parameter and initial stress on the propagation of the displacement waves u_1 and u_2 along the x-axis, the numerical results are presented graphically, as shown in **Figures 5** and **6**. These results are compared for different values of initial stress parameter and functionally graded parameter according to the following cases, A, B, C, and D,

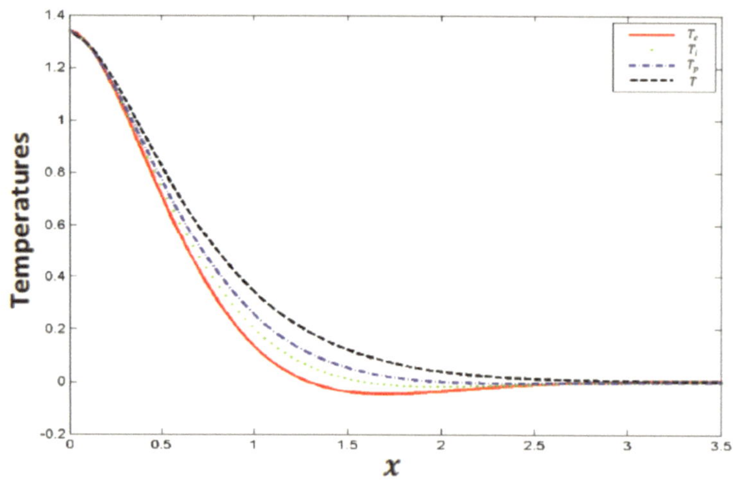

Figure 2.
Propagation of the temperature T_e, T_i, T_p and T waves along the x-axis.

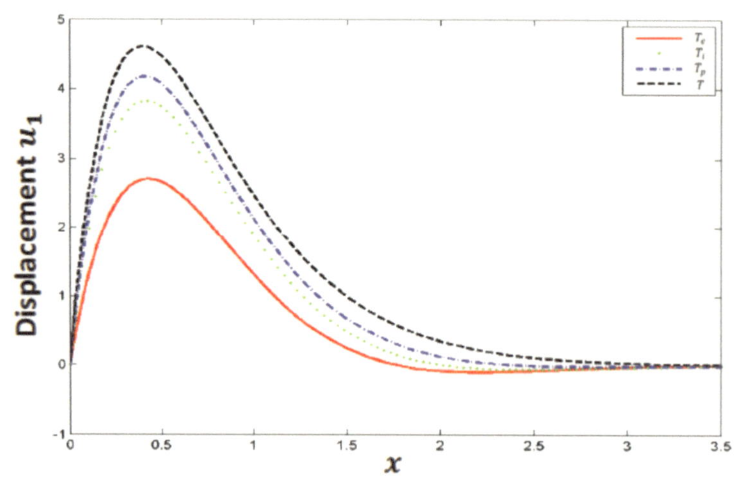

Figure 3.
Propagation of the displacement u_1 waves along the x-axis.

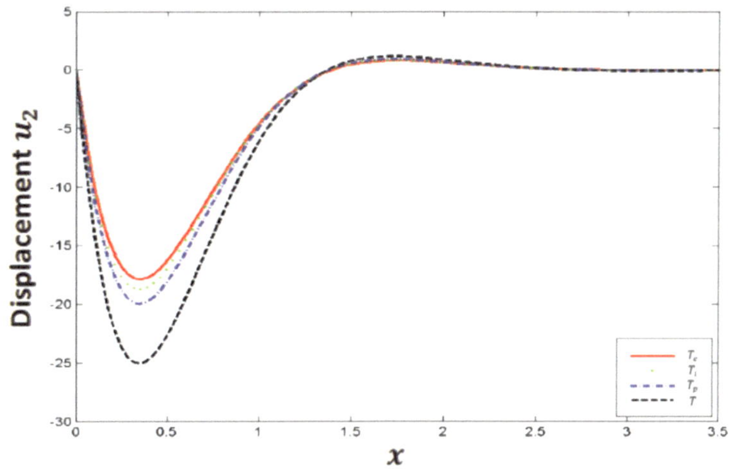

Figure 4.
Propagation of the displacement $\mathbf{u}_2$ *waves along the* $\mathbf{x}$*-axis.*

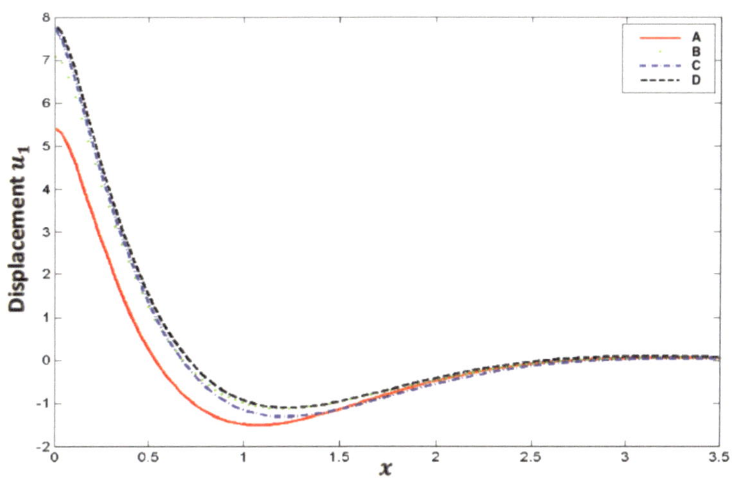

Figure 5.
Propagation of the displacement $\mathbf{u}_1$ *waves along the* $\mathbf{x}$*-axis.*

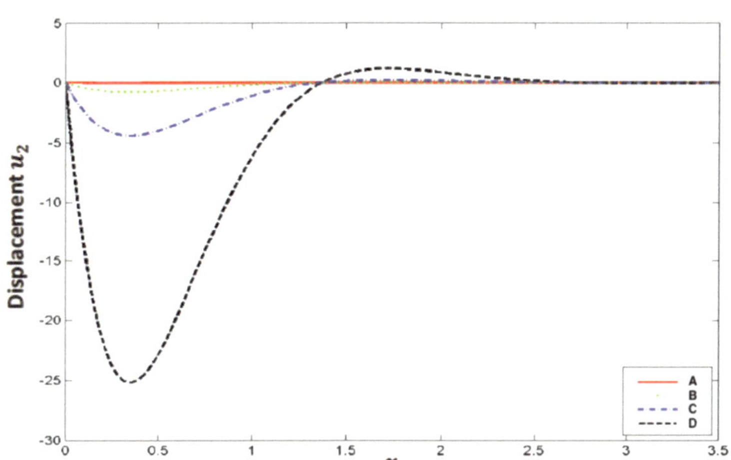

Figure 6.
Propagation of the displacement $\mathbf{u}_2$ *waves along the* $\mathbf{x}$*-axis.*

where A represents the numerical results for homogeneous ($m = 0$) structures in the absence of initial stress ($P = 0$), B represents the numerical results for functionally graded ($m = 0.5$) structures in the absence of initial stress ($P = 0$), C represents the numerical results for homogeneous ($m = 0$) structures in the presence of initial stress ($P = 0.5$), and D represents the numerical results for functionally graded ($m = 0.5$) structures in the presence of initial stress ($P = 0.5$). It can be seen from **Figures 5** and **6** that the effects of initial stress and functionally graded parameter are very pronounced.

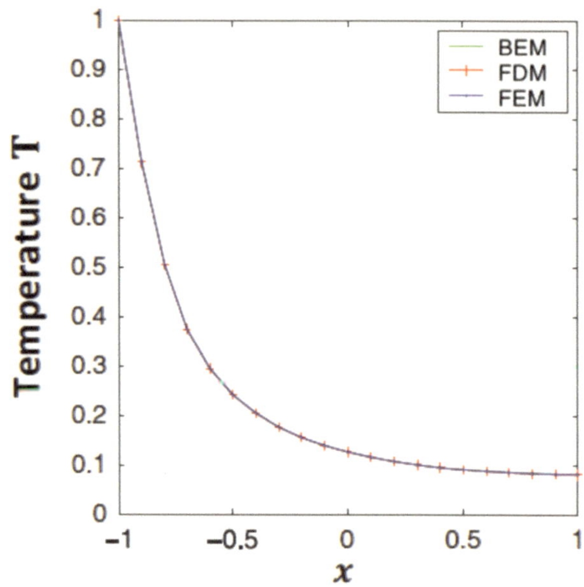

Figure 7.
Propagation of the temperature **T** *waves along the* **x**-*axis for BEM, FDM, and FEM.*

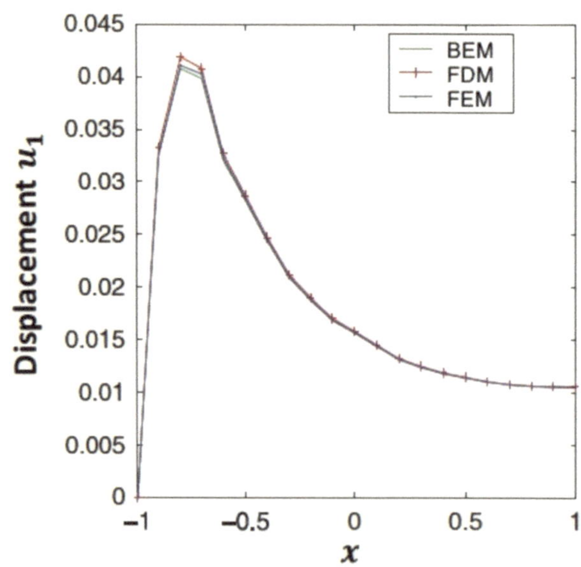

Figure 8.
Propagation of the displacement $\mathbf{u}_1$ *waves along the* **x**-*axis for BEM, FDM, and FEM.*

Since there are no available results for the three-temperature thermoelastic problem, except for Fahmy's research [10–14]. For comparison purposes with the special cases of other methods treated by other authors, we only considered a one-dimensional special case of nonlinear generalized magneto-thermoelastic of anisotropic structure [11, 12] as a special case of the considered problem. In the special case under consideration, the temperature and displacement wave propagation results are plotted in **Figures 7** and **8**. The validity and accuracy of our proposed BEM technique was demonstrated by comparing graphically the BEM results for the considered problem with those obtained using the finite difference method (FDM) of Pazera and Jędrysiak [68] and finite element method (FEM) of Xiong and Tian [69] results based on replacing heat conduction with three-temperature heat conduction; it can be noticed that the BEM results are found to agree very well with the FDM or FEM results.

6. Conclusion

Propagation of displacements and temperature acoustic waves in three-temperature nonlinear generalized magneto-thermoelastic ISMFGA structures has been studied, where we used the three-temperature nonlinear radiative heat conduction equations combined with electron, ion, and phonon temperatures. The BEM results of the considered model show the differences between electron, ion, phonon, and total temperature distributions within the ISMFGA structures. The effects of electron, ion, phonon, and total temperatures on the propagation of acoustic displacement waves have been investigated. Also, the effects of functionally graded parameter and initial stress on the propagation of acoustic displacement waves have been established. Since there are no available results for comparison, except for the one-temperature heat conduction problems, we considered the one-dimensional special case of our general model based on replacing three-temperature radiative heat conductions with one-temperature heat conduction for the verification and demonstration of the considered model results. In the considered special case, the BEM results have been compared graphically with the FDM and FEM, and it can be noticed that the BEM results are in excellent agreement with the FDM and FEM results.

Nowadays, knowledge and understanding of the propagation of acoustic waves of three-temperature nonlinear generalized magneto-thermoelasticity theory can be utilized by computer scientists and engineers, geotechnical and geothermal engineers, material science researchers and designers, and mechanical engineers for designing heat exchangers, semiconductor nanomaterials, and boiler tubes, as well as for chemists to observe the chemical reaction processes such as bond forming and bond breaking. In the application of three-temperature theories in advanced manufacturing technologies, with the development of soft machines and robotics in biomedical engineering and advanced manufacturing, acoustic displacement waves will be encountered more often where three-temperature nonlinear generalized magneto-thermoelasticity theory will turn out to be the best choice for thermomechanical analysis in the design and analysis of advanced ISMFGA structures using laser ultrasonics.

Author details

Mohamed Abdelsabour Fahmy
Faculty of Computers and Informatics, Suez Canal University, Ismailia, Egypt

*Address all correspondence to: mohamed_fahmy@ci.suez.edu.eg

References

[1] Telschow KL, Conant RJ. Optical and thermal parameter effects on laser generated ultrasound. The Journal of the Acoustical Society of America. 1990;**88**: 1494-1502

[2] Hurley DH, Spicer JB, Wagner JW, Murray TW. Investigation of the anisotropic nature of laser-generated ultrasound in zinc and unidirectional carbon epoxy composites. Ultrasonics. 1998;**36**:355-360

[3] Bernstein JR, Spicer JB. Line source representation for laser-generated ultrasound in aluminum. The Journal of the Acoustical Society of America. 2000;**107**:1352-1357

[4] White R. Generation of elastic waves by transient surface heating. Journal of Applied Physics. 1963;**34**: 3559-3567

[5] Scruby C, Dewhurst R, Hutchins D, Palmer S. Quantitative studies of thermally-generated elastic waves in laser irradiated metals. Journal of Applied Physics. 1980;**51**:6210-6216

[6] Rose LRF. Point-source representation for laser generated ultra-sound. The Journal of the Acoustical Society of America. 1984;**75**: 723-732

[7] Doyle P. On epicentral waveforms for laser-generated ultrasound. Journal of Physics D: Applied Physics. 1986;**19**: 1613-1623

[8] McDonald F. Practical quantitative theory of photoacoustic pulse generation. Applied Physics Letters. 1989;**54**:1504-1506

[9] Spicer J. Laser ultrasonics in finite structures: Comprehensive modeling with supporting experiment [PhD thesis]. The Johns Hopkins University; 1991

[10] Fahmy MA. A new boundary element strategy for modeling and simulation of three temperatures nonlinear generalized micropolar-magneto-thermoelastic wave propagation problems in FGA structures. Engineering Analysis with Boundary Elements. 2019;**108**:192-200

[11] Fahmy MA. A new computerized boundary element model for three-temperature nonlinear generalized thermoelastic stresses in anisotropic circular cylindrical plate structures. In: Awrejcewicz J, Grzelczyk D, editors. Dynamical Systems Theory. London, UK: IntechOpen; 2019. pp. 1-17

[12] Fahmy MA. Boundary element model for nonlinear fractional-order heat transfer in magneto-thermoelastic FGA structures involving three temperatures. In: Ebrahimi F, editor. Mechanics of Functionally Graded Materials and Structures. London, UK: IntechOpen; 2019. pp. 1-22

[13] Fahmy MA. Boundary element mathematical modelling and boundary element numerical techniques for optimization of micropolar thermoviscoelastic problems in solid deformable bodies. In: Sivasankaran S, Nayak PK, Günay E, editors. Mechanics of Solid Deformable Bodies. London, UK: IntechOpen; 2020. pp. 1-21

[14] Fahmy MA. Boundary element modeling and optimization based on fractional-order derivative for nonlinear generalized photo-thermoelastic stress wave propagation in three-temperature anisotropic semiconductor structures. In: Sadollah A, Sinha TS, editors. Recent Trends in Computational Intelligence. London, UK: IntechOpen; 2020. pp. 1-16

[15] Abd-Alla AM, El-Naggar AM, Fahmy MA. Magneto-thermoelastic

problem in non-homogeneous isotropic cylinder. Heat and Mass Transfer. 2003; **39**:625-629

[16] Hu Q, Zhao L. Domain decomposition preconditioners for the system generated by discontinuous Galerkin discretization of 2D-3T heat conduction equations. Communications in Computational Physics. 2017;**22**: 1069-1100

[17] Sharma N, Mahapatra TR, Panda SK. Thermoacoustic behavior of laminated composite curved panels using higher-order finite-boundary element model. International Journal of Applied Mechanics. 2018;**10**:1850017

[18] Fahmy MA. A time-stepping DRBEM for magneto-thermo-viscoelastic interactions in a rotating nonhomogeneous anisotropic solid. International Journal of Applied Mechanics. 2011;**3**:1-24

[19] Fahmy MA. A time-stepping DRBEM for the transient magneto-thermo-visco-elastic stresses in a rotating non-homogeneous anisotropic solid. Engineering Analysis with Boundary Elements. 2012;**36**: 335-345

[20] Fahmy MA. Numerical modeling of transient magneto-thermo-viscoelastic waves in a rotating nonhomogeneous anisotropic solid under initial stress. International Journal of Modeling. Simulation and Scientific Computing. 2012;**3**:1250002

[21] Fahmy MA. Transient magneto-thermo-viscoelastic stresses in a rotating nonhomogeneous anisotropic solid with and without a moving heat source. Journal of Engineering Physics and Thermophysics. 2012;**85**:950-958

[22] Fahmy MA. Transient magneto-thermo-elastic stresses in an anisotropic viscoelastic solid with and without moving heat source. Numerical Heat Transfer, Part A: Applications. 2012;**61**: 547-564

[23] Fahmy MA. Transient magneto-thermoviscoelastic plane waves in a non-homogeneous anisotropic thick strip subjected to a moving heat source. Applied Mathematical Modelling. 2012; **36**:4565-4578

[24] Fahmy MA. The effect of rotation and inhomogeneity on the transient magneto-thermoviscoelastic stresses in an anisotropic solid. ASME Journal of Applied Mechanics. 2012;**79**:1015

[25] Fahmy MA. Computerized Boundary Element Solutions for Thermoelastic Problems: Applications to Functionally Graded Anisotropic Structures. Saarbrücken: Lambert Academic Publishing (LAP); 2017

[26] Fahmy MA. Boundary Element Computation of Shape Sensitivity and Optimization: Applications to Functionally Graded Anisotropic Structures. Saarbrücken: Lambert Academic Publishing (LAP); 2017

[27] Fahmy MA. A time-stepping DRBEM for 3D anisotropic functionally graded piezoelectric structures under the influence of gravitational waves. In: Facing the Challenges in Structural Engineering, Sustainable Civil Infrastructures. Proceedings of the 1st GeoMEast International Congress and Exhibition (GeoMEast 2017); 15–19 July 2017; Sharm El Sheikh, Egypt. 2017. pp. 350-365

[28] Fahmy MA. 3D DRBEM modeling for rotating initially stressed anisotropic functionally graded piezoelectric plates. In: Proceedings of the 7th European Congress on Computational Methods in Applied Sciences and Engineering (ECCOMAS 2016); 5–10 June 2016; Crete Island, Greece. pp. 7640-7658

[29] Fahmy MA. Boundary element solution of 2D coupled problem in

anisotropic piezoelectric FGM plates. In: Proceedings of the 6th International Conference on Computational Methods for Coupled Problems in Science and Engineering (Coupled Problems 2015); 18–20 May 2015; Venice, Italy. pp. 382-391

[30] Fahmy MA. The DRBEM solution of the generalized magneto-thermo-viscoelastic problems in 3D anisotropic functionally graded solids. In: Proceedings of the 5th International Conference on Coupled Problems in Science and Engineering (Coupled Problems 2013); 17–19 June 2013; Ibiza, Spain. pp. 862-872

[31] Fahmy MA. A computerized boundary element model for simulation and optimization of fractional-order three temperatures nonlinear generalized piezothermoelastic problems based on genetic algorithm. In: AIP Conference Proceedings 2138 of Innovation and Analytics Conference and Exihibiton (IACE 2019); 25-28 March 2019; Sintok, Malaysia. p. 030015

[32] Soliman AH, Fahmy MA. Range of applying the boundary condition at fluid/porous Interface and evaluation of beavers and Joseph's slip coefficient using finite element method. Computation. 2020;**8**:14

[33] Eskandari AH, Baghani M, Sohrabpour S. A time-dependent finite element formulation for thick shape memory polymer beams considering shear effects. International Journal of Applied Mechanics. 2019;**10**:1850043

[34] Huang R, Zheng SJ, Liu ZS, Ng TY. Recent advances of the constitutive models of smart materials—Hydrogels and shape memory polymers. International Journal of Applied Mechanics. 2020;**12**:2050014

[35] Othman MIA, Khan A, Jahangir R, Jahangir A. Analysis on plane waves through magneto-thermoelastic microstretch rotating medium with temperature dependent elastic properties. Applied Mathematical Modelling. 2019;**65**:535-548

[36] Ezzat MA, El-Bary AA. Application of fractional order theory of magneto-thermoelasticity to an infinite perfect conducting body with a cylindrical cavity. Microsystem Technologies. 2017; **23**:2447-2458

[37] Fahmy MA. A new computerized boundary element algorithm for cancer modeling of cardiac anisotropy on the ECG simulation. Asian Journal of Research in Computer Science. 2018;**2**: 1-10

[38] Fahmy MA. Implicit-explicit time integration DRBEM for generalized magneto-thermoelasticity problems of rotating anisotropic viscoelastic functionally graded solids. Engineering Analysis with Boundary Elements. 2013; **37**:107-115

[39] Fahmy MA. Generalized magneto-thermo-viscoelastic problems of rotating functionally graded anisotropic plates by the dual reciprocity boundary element method. Journal of Thermal Stresses. 2013;**36**:1-20

[40] Fahmy MA. A three-dimensional generalized magneto-thermo-viscoelastic problem of a rotating functionally graded anisotropic solids with and without energy dissipation. Numerical Heat Transfer, Part A: Applications. 2013;**63**:713-733

[41] Fahmy MA. A 2-D DRBEM for generalized magneto-thermo-viscoelastic transient response of rotating functionally graded anisotropic thick strip. International Journal of Engineering and Technology Innovation. 2013;**3**:70-85

[42] Fahmy MA, Salem AM, Metwally MS, Rashid MM. Computer implementation of the DRBEM for

studying the generalized thermoelastic responses of functionally graded anisotropic rotating plates with one relaxation time. International Journal of Applied Science and Technology. 2013; **3**:130-140

[43] Fahmy MA, Salem AM, Metwally MS, Rashid MM. Computer implementation of the DRBEM for studying the classical uncoupled theory of thermoelasticity of functionally graded anisotropic rotating plates. International Journal of Engineering Research and Applications. 2013;**3**: 1146-1154

[44] Fahmy MA. A computerized DRBEM model for generalized magneto-thermo-visco-elastic stress waves in functionally graded anisotropic thin film/substrate structures. Latin American Journal of Solids and Structures. 2014;**11**:386-409

[45] Fahmy MA, Salem AM, Metwally MS, Rashid MM. Computer implementation of the DRBEM for studying the classical coupled thermoelastic responses of functionally graded anisotropic plates. Physical Science International Journal. 2014;**4**: 674-685

[46] Fahmy MA, Salem AM, Metwally MS, Rashid MM. Computer implementation of the DRBEM for studying the generalized thermo elastic responses of functionally graded anisotropic rotating plates with two relaxation times. British Journal of Mathematics & Computer Science. 2014;**4**:1010-1026

[47] Fahmy MA. A 2D time domain DRBEM computer model for magneto-thermoelastic coupled wave propagation problems. International Journal of Engineering and Technology Innovation. 2014;**4**:138-151

[48] Brebbia CA, Telles JCF, Wrobel L. Boundary Element Techniques in Engineering. New York: Springer-Verlag; 1984

[49] Wrobel LC, Brebbia CA. The dual reciprocity boundary element formulation for nonlinear diffusion problems. Computer Methods in Applied Mechanics and Engineering. 1987;**65**:147-164

[50] Partridge PW, Wrobel LC. The dual reciprocity boundary element method for spontaneous ignition. International Journal for Numerical Methods in Engineering. 1990;**30**:953-963

[51] Partridge PW, Brebbia CA. Computer implementation of the BEM dual reciprocity method for the solution of general field equations. Communications in Applied Numerical Methods. 1990;**6**:83-92

[52] Partridge PW, Brebbia CA, Wrobel LC. The Dual Reciprocity Boundary Element Method. Southampton: Computational Mechanics Publications; 1992

[53] Gaul L, Kögl M, Wagner M. Boundary Element Methods for Engineers and Scientists. Berlin: Springer-Verlag; 2003

[54] Fahmy MA, Al-Harbi SM, Al-Harbi BH. Implicit time-stepping DRBEM for design sensitivity analysis of magneto-thermo-elastic FGA structure under initial stress. American Journal of Mathematical and Computational Sciences. 2017;**2**:55-62

[55] Fahmy MA. The effect of anisotropy on the structure optimization using golden-section search algorithm based on BEM. Journal of Advances in Mathematics and Computer Science. 2017;**25**:1-18

[56] Fahmy MA. DRBEM sensitivity analysis and shape optimization of rotating magneto-thermo-viscoelastic FGA structures using golden-section

search algorithm based on uniform bicubic B-splines. Journal of Advances in Mathematics and Computer Science. 2017;**25**:1-20

[57] Fahmy MA. A predictor-corrector time-stepping DRBEM for shape design sensitivity and optimization of multilayer FGA structures. Transylvanian Review. 2017;**XXV**: 5369-5382

[58] Fahmy MA. Shape design sensitivity and optimization for two-temperature generalized magneto-thermoelastic problems using time-domain DRBEM. Journal of Thermal Stresses. 2018;**41**: 119-138

[59] Fahmy MA. Boundary element algorithm for modeling and simulation of dual-phase lag bioheat transfer and biomechanics of anisotropic soft tissues. International Journal of Applied Mechanics. 2018;**10**:1850108

[60] Fahmy MA. Shape design sensitivity and optimization of anisotropic functionally graded smart structures using bicubic B-splines DRBEM. Engineering Analysis with Boundary Elements. 2018;**87**:27-35

[61] Fahmy MA. Modeling and optimization of anisotropic viscoelastic porous structures using CQBEM and moving asymptotes algorithm. Arabian Journal for Science and Engineering. 2019;**44**:1671-1684

[62] Fahmy MA. Boundary element modeling and simulation of biothermomechanical behavior in anisotropic laser-induced tissue hyperthermia. Engineering Analysis with Boundary Elements. 2019;**101**: 156-164

[63] Fahmy MA, Al-Harbi SM, Al-Harbi BH, Sibih AM. A computerized boundary element algorithm for modeling and optimization of complex magneto-thermoelastic problems in MFGA structures. Journal of Engineering Research and Reports. 2019;**3**:1-13

[64] Fahmy MA. A new LRBFCM-GBEM modeling algorithm for general solution of time fractional order dual phase lag bioheat transfer problems in functionally graded tissues. Numerical Heat Transfer, Part A: Applications. 2019;**75**:616-626

[65] Fahmy MA. Design optimization for a simulation of rotating anisotropic viscoelastic porous structures using time-domain OQBEM. Mathematics and Computers in Simulation. 2019;**66**: 193-205

[66] Fahmy MA. A new convolution variational boundary element technique for design sensitivity analysis and topology optimization of anisotropic thermo-poroelastic structures. Arab Journal of Basic and Applied Sciences. 2020;**27**:1-12

[67] Zenkour A, Allehaibi AM, Fahmy MA. Three-temperature nonlinear generalized magneto-thermoelastic stresses in anisotropic cylindrical shells. Journal of Applied and Computational Mechanics. DOI: 10.22055/jacm.2020.33101.2151

[68] Pazera E, Jędrysiak J. Effect of microstructure in thermoelasticity problems of functionally graded laminates. Composite Structures. 2018; **202**:296-303

[69] Xiong QL, Tian XG. Generalized magneto-thermo-microstretch response during thermal shock. Latin American Journal of Solids and Structures. 2015; **12**:2562-2580

Chapter 8

Underwater Ambient Noise

Vijaya Baskar Veeriayan and Rajendran V.

Abstract

Underwater ambient noise is primarily a background noise which is a function of time, location, and depth. Background baseline of the noise in ocean is represented by ambient noise generated from the ocean surface due to wind and rain. This understanding pertained to ambient noise under various conditions will help in improving the signal-to-noise ratio (SNR) of marine instruments. It is of prime importance to detect the signals such as sound of a submarine or echo from a target surpassing this ambient noise. Ambient noise excludes all forms of self noise, such as the noise of current flow around the measurement hydrophone and its supporting structure. It should also exclude all forms of electrical noise. It is also defined as the residual noise that remains after all easily identifiable sound sources are eliminated. In the absence of sound from ships and marine life, underwater ambient noise levels (NL) are dependent mainly on wind speeds at frequencies between 100 Hz and 25 KHz.

Keywords: ambient noise, underwater, shallow water, wind noise, background noise

1. Introduction

Nature is wonderful to observe, study, or analyze its diversity. It had thrown challenges to man in the past, and still this study is ever exploring. For the past many years, acoustics and oceanographic study have been interrelated. Ocean is a pandemonium of different sounds. Underwater acoustics has become the natural interest to many researchers throughout the world for the simple reason that it is so complex to study and analyze. The sources of sound are breaking waves, marine life, various sources of nature, and rain. Shipping also becomes an important factor for the noise generation along with several other man-made sources like military sonar for the source of sound. These sounds exist throughout the year along with the sounds that are seasonal.

Underwater ambient noise is a constituent of background noise that depends on depth, time, and also w.r.t location. Self noise does not belong to the category of ambient noise [1]. Ambient noise is the residual noise that would exist even after recognizable sources of sound are removed. Ship traffic contributes to the major component of ambient noise, whereas the noise generated by a nearby ship is considered as sound signal and is not a part of ambient noise [2].

For valid measurements of ambient noise, all possible sources of self-noise must be eliminated or at least reduced to an insignificant contribution to the total noise level (NL). Self noise sources, such as cable strumming, splashes of waves against the hydrophone cable, and sometimes even crabs crawling on the hydrophone, must be absent.

IntechOpen

2. Ambient noise: frequency dependency

The classification of primary sources of noise can be done by taking frequency as a parameter. For example, ambient noise because of distant ship traffic exists between 20 and 500 Hz. In the absence of noise due to ships that are nearer to the receiver, also noise can be identified, which is obviously due to distant ships. Noise is higher in high ship traffic locations. For example, in the southern hemisphere, where there exist fewer ships, naturally low-frequency ambient noise levels are recorded. With the increased ship traffic, the noise produced is also high in international waters [3].

Ambient noise generated due to spray and bubbles related to breaking waves fall in the range of 500–100,000 Hz [4, 5]. With the increase in wind speed, ambient noise also increases. After 100,000 Hz, thermal noise dominates. Thermal noise is the noise due to random motion of water molecules, which is sometimes considered as threshold [6].

Knudsen et al. [7] studied ambient noise in harbors and in coastal waters. This study reveals the fact that the noise level raises with the increase in speed of the wind and wave height. At a specified wind speed, noise level decreases along with the incremental values of frequency. It has been observed that in the frequency range of 500 Hz to 50 KHz, ambient noise is a function of wind speed. This encourages researchers to use this as a means to measure wind speed over the ocean.

Apart from the sources of ambient noise that have been listed out, several intermittent sources exist in the ocean, which include marine life, man-made, and natural processes. These intermittent sources are limited to particular regions of ocean.

3. Ambient noise due to intermittent sources

The influence of intermittent sources on ambient noise is significant, which includes various factors such as marine life, rain noise, shipping noise, etc.

3.1 Marine life

A distinct and innumerous variety of sounds are produced by marine life. Humpback whales, dolphins, and other marine mammals generate sounds of a broad range of frequencies, and these sounds can be treated as infrasonic or ultrasonic. Specific categories of fish and marine invertebrates are also sources of sound. For example, special species of fish includes toadfish and croaker, and some of the marine invertebrates are snapping shrimp. Information pertained to surroundings is obtained by marine animals using sound. This dependency on sound extends in communicating, navigating, and even feeding. For example, dolphin depends on sound to search food and smell enemies. This is done by transmitting sound pulses for short intervals, and then with the echo received, dolphins detect objects in sea. Most of the time, this kind of communication contributes to raise in ambient noise levels for a maximum value of 25 dB. In the case of whales, noise levels are up to 190 dB at 10–25 Hz frequencies [6].

Sound produced by snapping shrimp inhabiting in shallow waters contributes to background noise. In general, prevalence of this can be seen in semitropical and shallow tropical waters which contain seabed with rock and shell, as they provide the necessary concealment. Colonies of snapping shrimp produce sounds at frequencies of 2–15 KHz. Individually these have peak-to-peak source levels of 189 underwater dB at 1 m.

It is also possible that the sounds that are known to humans and pertained to underwater have the sources that include more than the species that have been identified.

3.2 Natural physical processes

Most of the physical processes produce higher levels of sounds with source levels up to 260 underwater dB at 1 m. Rain, undersea earthquakes, and underwater volcanoes generate intermittent sounds in the ocean. Heavy rain raises noise level by 35 underwater dB from few Hz to beyond 20 KHz.

3.3 Human activities

Human activity is one of the important sources of background noise. Sounds in underwater are used for many purposes like communication, navigation, and fishing, and the point of observation is that one act is source for the other activity. Several by-products can also be seen like noise generation by offshore activities such as oil drilling, manufacturing, and other industrial acoustic pollutions. Incidentally and evidently, all of these factors, more or less, have a human hand in it. So, it is very clear that human activities also contribute to ambient noise of the underwater acoustics.

3.4 Rain noise

Nicolaas Bom [8] studied the influence of rain on undersea noise level. Experiments were conducted in Italy at the lake of Sarzana near La Spezia. Data is analyzed between 300 Hz and 96 KHz. The observation and inference of this work reflected the effect of rain on noise, i.e., noise level increases w.r.t to increased rainfall rate. It was concluded that "it corresponded to higher frequency energy."

In 1986, Nystuen [9] reported that a light rain with total rainfall of <2.5 mm per hour produced a spectral peak at 15 KHz (broadband wind sensitive). And, a heavy rain where a total rainfall was more than 7.5 mm per hour produced a wideband noise ranging from 4 to 21 KHz. A similar type of study and data analysis was done by Deane [10] for a period of 1 year, and this work also revealed the effect of rain noise.

3.5 Shipping noise

Shipping noise can exhibit both spatial and temporal variability. The spatial variability is largely governed by the distribution of shipping routes in the oceans. Temporal variability can be due to the seasonal activities of fishing fleets. The noise generated by coastal shipping and by high-latitude shipping can contribute to the noise field in the deep sound channel in tropical and subtropical ocean areas. Specifically, coastal shipping noise is intruded into the deep sound channel through the process of downslope conversion. High-latitude shipping noise is generated through the latitudinal dependence of the depth of the sound channel axis.

3.6 Wind-generated noise

Noise due to wind is a major contributor of the total ambient noise. Wind-associated ambient noise was predicted and analyzed earlier by many researchers. Ambient noise depends on varying wind speeds over the frequency range of 500 Hz to 50 KHz. In this range of frequency, i.e., in KHz, hydrophone is used as an anemometer [1]. Wind speed is considered as the major constituent of ambient noise in the absence of ships and tumid marine life.

Wind noise can be treated as a typical case of random noise. Wind noise in the frequency range of 500 Hz to 20 KHz is called as Knudsen noise, because Knudsen discovered that it correlated very well with wind speed [7].

4. Ambient noise level

Wenz curves are very helpful in the prediction of the level of ambient noise under given constraints. Wenz curves as shown in **Figure 1** represent the average ambient noise for different shipping traffic levels and wind speeds.

Ambient noise level variation w.r.t frequency can be well understood from Wenz curves, and this can be summarized as follows:

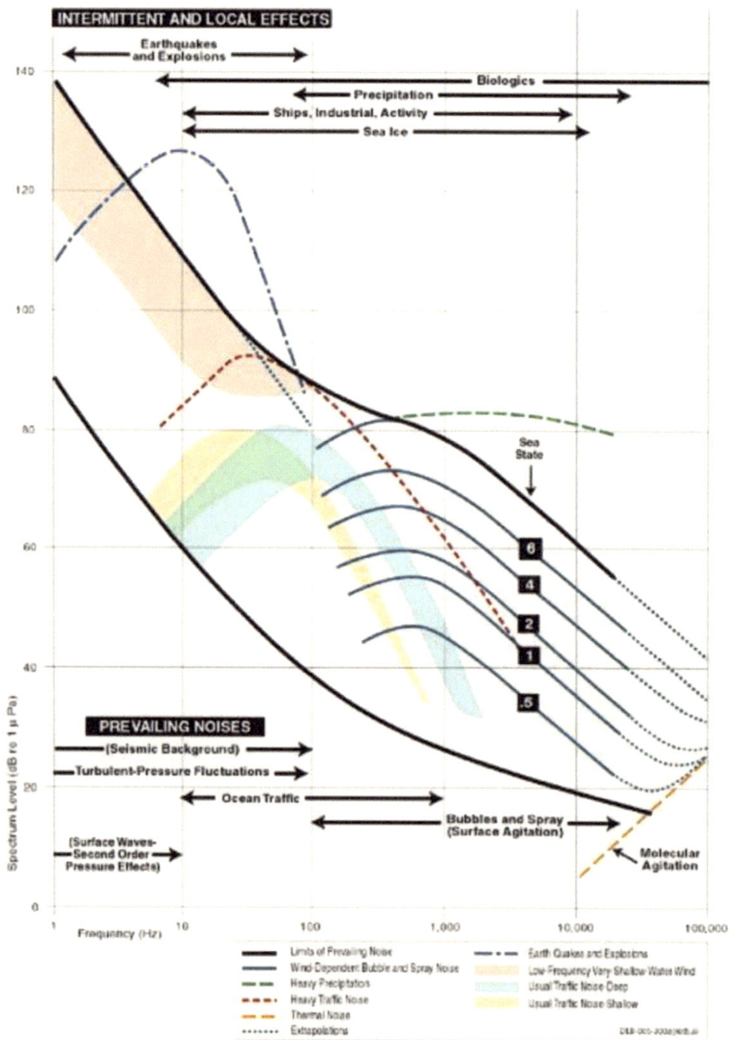

Figure 1.
Wenz curve [11].

- 10–100 Hz—In this band noise level depends mostly on shipping density and activities related to industries. In such cases noise levels are in the range of 60–90 dB with less-frequency dependence.

- 100–1000 Hz—Here shipping is the primary source of noise. Sea surface agitation is also a contributing factor to noise with this frequency range.

- 1–100 kHz—Sea surface agitiation is the dominant factor unless and otherwise marine mammals or rain is present.

- >100 kHz—Noise in this range is dominated by electronic and thermal noise.

The main inference can be drawn out of Wenz curves as "noise level (NL) decreases as depth increases and also w.r.t change in frequency, because most noise sources are relatively closer to the surface."

5. Ambient noise in shallow water

Shallow water ambient noise is highly random due to the waveguide nature of the environment and bottom reflection [12]. There are two definitions of shallow water: hypsometric and acoustic. The hypsometric definition is based on the fact that most continents have continental shelves bordered by the 200 m bathymetric contour, beyond which the bottom generally falls off rapidly into deep water. Therefore, shallow water is often taken to mean continental shelf waters shallower than 200 m. Using this definition, shallow water represents about 7.5 percent of the total ocean area.

Acoustically, shallow water conditions exist whenever the propagation is characterized by numerous encounters with both the sea surface and the seafloor. By this definition, some hypsometrically shallow water areas are acoustically deep. Alternatively, the deep ocean may be considered shallow when low-frequency and long-range propagation conditions are achieved through repeated interactions with the sea surface and the seafloor.

Shallow water regions are distinguished from deep water regions by the relatively greater role played in shallow water by the reflecting and scattering boundaries. Also differences from one shallow water region to another are primarily driven by differences in the structure and composition of the seafloor. Apart from water depth, the seafloor is perhaps the most important part of the marine environment that distinguishes shallow water propagation from deep water propagation. The most common shallow water bottom sediments are sand, silt, and mud, with compressional sound speeds greater than that of the overlying water.

In shallow water, in the absence of local shipping and biological noise, wind noise dominates the noise of distant shipping over the entire frequency range [1]. The reason for this is that the deep favorable propagation paths traveled by distant shipping noise in deep water are absent in shallow water.

Shallow water is known for its variability. Waters close to shore and in busy harbors are dynamic locations, where rapid noise changes take place. Even though shallow water has the variability as a notable feature due to variable background of biological and shipping activities, at high frequencies and even at high wind speeds, this level is significantly constant from location to location at the same wind speed. Although noise field characterization is complex in shallow waters, it has to be carried out because of its greater applications in naval operations.

Ambient noise is more in shallow water as the noise is pinned between the seafloor and the surface of the ocean. In shallow water (depth of 5–200 m), acoustic systems like sonar, echo sounder, and sub-bottom profiler suffer a huge loss due to ambient noise [13]. Wind speed in shallow water may vary from 0.5 m/s to 300 m/s; hence it causes disturbance to surface water, and these disturbances thereby propagate towards the seabed. Due to these factors the SNR value reduces. This poses a greater challenge to developing a method or algorithm to improve the SNR value. Hence, to improve the effectiveness of underwater acoustic instruments, the measurement and characterization of ambient noise is more significant.

6. Characterization and denoising of ambient noise

Characterization of ambient noise had become the epicenter during World War II. The need to develop acoustic sensor and systems had become of prime importance for processing different sound signatures of submarines [1]. This emphasized the need of noise characterization, and with this interest, many engineers and researchers started working with sensors and sonar for signal reception and processing.

After characterization of ambient noise, data analysis comes, which is essential for research and pragmatic approach. It serves two objectives: determining parameters that are needed for constructing necessary model and thereafter authenticating the constructed model.

Irrespective of physical measurements or quantitative modeling, data has one or more issues to deal with: (1) shorter total data span, (2) nonstationary data, and (3) data representing nonlinear processes. Here the first two issues are interrelated, for section of data shorter than the longest timescale of a stationary process may appear to be nonstationary. There are limited options available to analyze this kind of data (http://rcada.ncu.edu.tw/).

The recovery of the signal buried in ambient noise is of importance for the target's signal detection, recognition, and classification at low signal-to-noise ratio (Tieshuang et al. [14]).

7. Conclusion

In this chapter we have done a detailed study about underwater ambient noise, various sources that produce the noise, and shallow water ambient noise. In science and engineering, noise is defined as an unwanted energy. Noise is relentless that blots out or reduces the intensity of a signal. Noise can be a disturbance that may be internally generated because of the process itself or due to some irresolvable issues or may be due to intermittent and local instability. This can be produced by various recording systems and different types of sensors. It is important to analyze the data differences that would exist between actual data and the noise so that attempts can be done to remove this noise.

Author details

Vijaya Baskar Veeriayan[1*] and Rajendran V.[2*]

1 School of EEE, Sathyabama Institute of Science and Technology, Chennai, India

2 Department of ECE, VELS University, Chennai, India

*Address all correspondence to: vijayabaskar.ece@sathyabama.ac.in
and drvrajen@gmail.com

References

[1] Urick RJ. Ambient Noise in the Sea. Peninsula Publishing; 1984

[2] Hodges RP. False Alarms, False Contacts, and False Targets, in Underwater Acoustics: Analysis, Design and Performance of Sonar. Chichester, UK: John Wiley & Sons Ltd; 2010

[3] Andrew RK, Howe BM, Mercer JA, Dzieciuch MA. Ocean ambient sound: Comparing the 1960s with the 1990s for a receiver off the California coast. ARLO. 2002;**3**:65-70

[4] Franz GJ. Splashes as sources of sound in liquids. The Journal of the Acoustical Society of America. 1959;**31**:1080

[5] Medwin H, Beaky MM. Bubble sources of the Knudsen Sea noise spectra. The Journal of the Acoustical Society of America. 1989;**86**:1124-1130

[6] Available from: http://www.dosits.org/science/soundsinthesea/commonsounds/

[7] Knudsen VO, Alford RS, Emling JW. Underwater ambient noise. Journal of Marine Research. 1948;**7**:410-429

[8] Bom N. Effect of rain on underwater noise level. The Journal of the Acoustical Society of America. 1969;**45**:296

[9] Nystuen JA. Acoustical rainfall analysis: Rainfall drop size distribution using the underwater sound field. Journal of Atmospheric and Oceanic Tecnology. 1996;**13**:74-84

[10] Deane GB. Long time-base observations of surf noise. The Journal of the Acoustical Society of America. 2000;**107**:700-758

[11] Wenz GM. Acoustic ambient noise in the ocean: Spectra and sources. Journal of Acoustic Society of America. 1962;**34**:1936-1956

[12] Yang TC, Yoo K. Modeling the environmental influence on the vertical directionality of ambient noise in shallow water. The Journal of the Acoustical Society of America. 1997;**101**:2541

[13] Morris GB. Depth dependence of ambient noise in the northeastern Pacific Ocean. The Journal of the Acoustical Society of America. 1978;**64**:581

[14] Tieshuang H, Jinglin X, Peng H, Jie S. Research of ship-radiated signal denoising based on the DT CWT statistical wavelet model. In: 9th International Conference on Signal Processing, 2008 (ICSP 2008). 2008. pp. 2851-2854